Adaptive and Fault-Tolerant Control of Underactuated Nonlinear Systems

AUTOMATION AND CONTROL ENGINEERING SERIES

Adaptive and Fault-Tolerant Control of Underactuated Nonlinear Systems

Jiangshuai Huang • Yong-Duan Song

CRC Press
Taylor & Francis Group
Boca Raton London New York

CRC Press is an imprint of the
Taylor & Francis Group, an **informa** business

CRC Press
Taylor & Francis Group
6000 Broken Sound Parkway NW, Suite 300
Boca Raton, FL 33487-2742

First issued in paperback 2022

© 2018 by Taylor & Francis Group, LLC
CRC Press is an imprint of Taylor & Francis Group, an Informa business

No claim to original U.S. Government works

ISBN-13: 978-1-138-08902-0 (hbk)
ISBN-13: 978-1-03-233926-9 (pbk)
DOI: 10.1201/9781315109497

**Visit the Taylor & Francis Web site at
http://www.taylorandfrancis.com**

**and the CRC Press Web site at
http://www.crcpress.com**

To my parents, Li and Xiangjun
J. Huang

To my family
Y.-D. Song

Preface

Nonholonomic underactuated mechanical systems are special systems full of research interest and practical sense. Referring to those mechanical systems whose number of control variables is less then the degrees of freedom, the underactuated systems are abundant in real life, ranging from landing vehicles, surface ships, and underwater vehicles to spacecrafts. Mobile robots and surface vessels are two typical nonholonomic underactuated systems that receive tremendous consideration in literature. For the tracking and stabilization control of underactuated mechanical systems, many methodologies have been proposed for controlling these systems by researchers. However, there are still a lot of open problems in this area. Among them the tracking control with input saturation and output-feedback control in the presence of parametric uncertainties are two important problems to be solved in this thesis.

For tracking and stabilization control of nonholonomic mobile robots with input saturation, a new adaptive scheme is proposed to ensure that the bounds of the control torques are functions of only design parameters and reference trajectory. Thus suitable design parameters can be chosen so that such bounds are within the given saturation limits. For adaptive output-feedback tracking control of nonholonomic mobile robots, a new adaptive control scheme is proposed including designing a new adaptive state feedback controller and two high-gain observers to estimate the unknown linear and angular velocities respectively. For tracking control of underactuated ships we design the yaw axis torque in such a way that its corresponding subsystem is finite time stable, which makes it decouple from the second subsystem after a finite time. This enables us to design the torque in the surge axis independently. For adaptive output-feedback tracking control of underactuated ships, by using the prescribed performance bound technique, the position error and orientation error can be guaranteed converging to arbitrarily small residual sets at a pre-specified exponential rate.

The major difficulties of adaptive output feedback for nonholonomic mobile robots and underactuated ships are caused by simultaneous existence of nonholonomic constraints, unknown system parameters and a quadratic term of unmeasurable states in the mobile robot dynamic system as well as their couplings. To overcome these difficulties, a new adaptive control scheme is proposed including designing a new adaptive state feedback controller and two high-gain observers to estimate the unknown linear and angular velocities respectively. It is shown that the closed loop adaptive system is stable and the tracking errors are guaranteed to be within the arbitrarily small pre-specified bounds. The stability of the system is analyzed and established by using the Lyapunov approach. Due to the disturbance rejection property of the high-gain observers, it is shown that the state estimation errors are of order $O(\varepsilon)$, where ε is a small design parameter, and the tracking errors can be made arbitrarily small. With our proposed control scheme, a solution is now presented for the outstanding problem of adaptive output-feedback control for nonholonomic mobile robots and underactuated ships.

A vital problem in the control of underactuated ships is that failures occurring on the actuators may cause the control ineffectiveness and even lead to disasters. Such failures are often uncertain in time, value and pattern, and may cause instability or even catastrophic accidents. For the sake of reliability and safety, actuator failure compensation has received an increasing amount of attention and considerable achievements have been made based on various approaches. For surface ships, the actuator failure may have two different kinds of patterns. For one, the sway actuator is missing and there is redundancy in the surge and yaw actuator. In this case we shall develop a fault-tolerant control scheme for surge and yaw thrusters such that even if unknown failures happen to these thrusters, the control of underactuated ships could still be finished and control performance could be guaranteed. For another, the sway thruster does not have any redundancy and may suffer from total actuator failure. Then a fully actuated ship turns into an underactuated ship. In this case, normally the control scheme developed for a fully actuated ship could not be used anymore, since the control target changes. To avoid the switching of the control schemes, we have developed a unified actuated/underactuated control scheme for surface vessels, such that the sway thruster could be turned off actively, or passively due to actuator failure. In both cases, the control scheme does not need to be switched.

Because of its widespread potential applications in various fields such as mobile robot networks, intelligent transportation management, surveillance and monitoring, distributed coordination of multiple dynamic subsystems (also known as multi-agent systems) has achieved rapid development during the past decades. Formation control and flocking control are the most popular topics in this area; these have received significant attention by numerous researchers. This research aims at achieving an agreement for certain variables, such as the states or the outputs, of the subsystems in a group. A large number of effective control approaches have been proposed to solve the constant consensus problems and the non-constant consensus issues of tracking time-varying trajectories. We here focus on the formation and flocking control of a group of underactuated mechanical systems, and investigate new control schemes to overcome the difficulties in the formation and flocking of multiple underactuated systems.

Chongqing, China,   *Jiangshuai Huang*
DEC 2016   *Yongduan Song*

Contents

Part I Adaptive Control of Underactuated Nonlinear Systems with Input Saturation

List of Figures

List of Tables

Symbols and Acronyms

Algebraic Operators

A^T	Transpose of matrix A
A^{-1}	Inverse of matrix A
$\det(A)$	Determinant of matrix A
$\rho(A)$	Spectral radius of matrix A
$\mathrm{diag}(a_1, \ldots, a_n)$	Diagonal matrix with main given diagonal numbers
$\|\cdot\|$	ℓ^2 norm for vector or the induced norm for matrix
$\|\cdot\|_\infty$	ℓ^∞ norm for vector or the induced norm for matrix
$M > 0 \ (M \geq 0)$	M is positive definite(semi-definite)
$M_1 > M_2 \ (M_1 \geq M_2)$	$M_1 - M_2$ is positive definite(semi-definite)
$A \otimes B$	Kronecker product of A and B

Sets

$\mathbb{R}$ Set of real numbers
$\mathbb{Z}$ Set of integers
$\mathbb{N}$ Set of nonnegative integers

Others

$\mathbf{0}$	Zero vector with a compatible dimension
$\mathbf{1}$	Vector with a compatible dimension and all elements of one
$\lambda_{\max}(P)$	maximum eigenvalue of matrix P
$\lambda_{\min}(P)$	minimum eigenvalue of matrix P

1

Introduction

1.1 Underactuated Mechanical Systems

In the past few decades, the control of underactuated mechanical systems has been widely investigated in the control community. Underactuated mechanical systems refer to those mechanical systems which have more degree of freedom than the number of actuators. For example an unicycle mobile robot, which has two input signals while the degrees of freedom is three, i.e., the positions in X and Y coordinates, and its heading angle. Underactuated mechanical systems appear almost everywhere in real life, e.g., robotics, aerospace systems, surface marine systems, underwater marine systems, helicopters, road vehicles, etc. In general, the linear approximation approach is not applicable to this type of systems. Therefore controller design for underactutated mechanical systems is a challenging task, and it has both theoretical interest and practical significance. The reasons of "underactuation" are mainly due to the following reasons: 1) kinematics or dynamics of the system are inherently underactuated, e.g. nonholonomic mobile robots, aircraft, spacecraft; 2) part of the actuators are removed for reducing cost or other practical purposes, e.g. surface vessel without sway force or satellite with two thrusters; 3) actuator failure; 4) artificially creating a complex low-order nonlinear system to obtain insight into high-order underactuated systems, e.g. inverted pendulum, Beam-and-Ball system, etc.

1.2 Nonholonomic Constraints

Most of the underactuated mechanical systems are suffered from nonholonomic constraints. Let $q \in Q$ be the configuration vector of a mechanical system, then nonholonomic constraint can be defined as a set of non-integrable conditions satisfied by certain functions. For example, for the condition that $\phi(q)\dot{q} = 0$, there does not exist a $\Phi(q) = 0$ such that $d\Phi(q) = \phi(q)\dot{q}$. In classical mechanical systems, nonholonomic constraints can be divided into two classes, the first order nonholonomic constraints and the second order nonholonomic constraints. For the first order nonholonomic constraints, the constraints are defined on the coordinates and velocities of the mechanical systems, whereas for the second order nonholonomic constraints, the constraints involve accelerations of the mechanical systems. The first-order nonholonomic constraint is widely found in land vehicles like mobile robots or tractor and trailer systems, while the second order nonholonomic condition is normally required by surface marine vehicles, under-water vehicles and space vehicles like ships and spacecraft.

1.2.1 First-order Nonholonomic Constraint

In order to illustrate the first order nonholonomic constraint, we show an example of a nonholonomic mobile robot by considering a two-wheeled mobile robot with its kinematic model is given by

$$\dot{x} = v \cos \theta$$
$$\dot{y} = v \sin \theta$$
$$\dot{\theta} = w$$

where the forward velocity v and angular velocity w are inputs that can be controlled independently and it is assumed the front castor wheel and the rear wheels just roll without slipping, (x, y) denotes the coordinates of the center of mass and θ denotes the angle between the heading direction and x-axis. Let the center-line be the line in the middle of two actuated wheels and orthogonal to the line connecting two wheels. Then the rolling without slipping condition of mobile robot requires that the velocity orthogonal to the center-line is equal to zero, which means

$$\dot{x} \sin \theta - \dot{y} \cos \theta = 0 \tag{1.1}$$

Constraint (1.1) can not be integrated, i.e., it cannot be written as a time-derivative of some function of the state, i.e. $G(x, y, \theta)$. Therefore it is called a nonholonomic constraint. In fact, it can be shown that a first-order nonholonomic system cannot be stabilized by any smooth time-invariant static state-feedback based on Brockett's necessary lemma as in [1].

1.2.2 Second-order Nonholonomic Constraint

We consider the general form of an underactuated mechanical system with $q = (q_1, q_2, ..q_n)$ as follows:

$$M_{11}(q)\ddot{q}_i + M_{12}(q)\ddot{q}_j + C_1(q, \dot{q}) = F(q)u \tag{1.2a}$$

$$M_{21}(q)\ddot{q}_i + M_{22}(q)\ddot{q}_j + C_2(q, \dot{q}) = 0 \tag{1.2b}$$

We partition the state space configuration into two parts, q_i and q_j, where $q_i \in R^m$ denotes the subsystem actuated directly by control $u \in R^m$, and $q_j \in R^{n-m}$ denotes the subsystem that is not directly actuated by actuators leaving q_j as an un-actuated subsystem. Actually if there is no nontrivial function $\Psi(t, q, \dot{q})$ with $d\Psi/dt = 0$ being the solution to (1.2b), then (1.2b) becomes a nonholonomic constraint.

Typical examples of underactuated nonholonomic systems with second order nonholonomic constraints include robot manipulators, autonomous underwater vehicles, underactuated surface vehicles, acrobot systems, and vertical take-off and land spacecraft and aircraft. In contrast to systems with first-order nonholonomic constraints, the systems with second-order nonholonomic constraints have drift-terms which make the controller design more difficult. Such systems also do not satisfy Brockett's necessary lemma, which means there is no time-invariant state feedback controller that makes the systems asymptotically stable. As an example, consider the dynamics of an underactuated ship, which can be described as

$$\dot{\eta} = J(\eta)\nu \tag{1.3}$$

$$M\dot{\nu} + C(\eta, \nu)\nu + D(\nu)\nu = \begin{bmatrix} \tau \\ 0 \end{bmatrix} \tag{1.4}$$

where $\eta = [x, y, \phi]^T$ denotes the position and the orientation of the ship in the body-fixed frame, $\nu = [u, v, r]^T$ denotes the linear and angular velocities of the ship, M is the inertia matrix, $C(\eta, \nu)$ denotes the Coriolis and centripetal matrix, $D(\nu)$ is the damping matrix.

1.2.3 Literature Review of Underactuated Nonlinear Systems

Typical examples of underactuated mechanical systems are nonholonomic mobile robots and underactuated ships, which have received lots of attention from the control community. In control theory, stabilization is usually regarded as a special case of the tracking problem. However for controlling underactuated mechanical systems with nonholonomic constraints such as nonholonomic mobile robots, the stabilization and tracking problems are totally different, and thus they are normally considered separately.

For nonholonomic mobile robots, due to the Brockett's necessary condition [1], there exists no continuous time-invariant state-feedback controller that asymptotically stabilizes the mobile robot. For the stabilization problem, the past abundant literature aims at developing a suitable discontinuous time-invariant stabilizer [3] or time-varying stabilizer [2] or hybrid stabilizers [4, 5]. The seminal work of Samson [2] introduced the first time-varying feedback stabilizer for a wheeled mobile robot. For tracking control, a time-varying state-feedback tracking controller [6] is proposed for kinematics and simplified dynamics of mobile robots based on backstepping [8]. In [7], an adaptive tracking controller is designed for nonholonomic mobile robots with parametric uncertainties. In [9] by using coordinate transformation, an adaptive controller is designed for nonholonomic mobile robots such that the tracking and stabilization problems can be solved with only one controller. In [10] a global time-varying output feedback controller that can solve tracking and stabilization for nonholonomic mobile robots is proposed based on a coordinate transformation. In [14], a transverse function approach is proposed for the stabilization of arbitrary reference trajectories for nonholonomic mobile robots based on a change of coordinates.

For the underactuated ships, in [16], the method developed for chained-form systems is applied for tracking control of underactuated ships through a coordinate transformation. Exponential stability of the closed-loop system is ensured and the position and the angle of the ship are steered to track the reference trajectory. In [17], two tracking control schemes are proposed using Lyapunov's direct and passivity approaches for underactuated ships. In [18] a tracking controller is developed for underactuated surface ships to globally asymptotically track a reference which is allowed to be a curve including a straight line. In [19] an universal controller is proposed for underactuated ships to achieve stabilization and tracking control simultaneously based on Lyapunov's direct method and backstepping technique. In [20], a continuous time-varying tracking controller is proposed to yield globally bounded tracking by transforming the ship error system into a skew-symmetric form and generating a time-varying dynamic oscillator. In [21] a global robust adaptive controller is proposed for an underactuated ship to follow a reference path under both constant and time-varying disturbances and parametric uncertainties. In [22] a global smooth controller is proposed for underactuated ships that achieves the practical stabilization of arbitrary reference trajectories including fixed points and non-admissible trajectories based on the transverse function approach developed in [14]. In [23] a nonlinear model-based adaptive output feedback controller is proposed for global asymptotic tracking in the presence of parametric uncertainties associated with nonlinear ship dynamics. In [25] by nonlinear coordinate changes to transform the ship dynamics to a system affine in the ship's velocities, observers are designed to estimate unmeasured velocities, and tracking controllers are then proposed based on Lyapunov's direct method and backstepping.

For general underactuated mechanical systems, several researchers have proposed controllers to achieve stabilization and tracking objectives. In [26] a sliding mode control scheme which robustly stabilizes a class of underactuated mechanical systems with parametric uncertainties that are not linearly controllable is proposed. In [27] a passivity based interconnection and damping assignment control scheme is proposed to address the stabilization of underactuated mechanical systems by modifying both the potential and kinetic energies. In [28] cascade normal forms including strict feedback form, feedforward form and nontriangular quadratic form are introduced for underactuated mechanical systems to facilitate control design. In [29] a hybrid switching control strategy that combines switching control, proportional control and derivative control is proposed for underactuated mechanical systems.

In [30] the controllability and stabilizability results are derived for a class of underactuated mechanical systems characterized by nonintegrable dynamics relations.

1.3 Motivations and Control Objectives

In practical applications, the inputs of the underactuated mechanical systems like mobile robots and underactuated ships are provided by the motors, whose outputs can not be unlimited. Thus when we design controllers for the underactuated mechanical systems the input saturation problem must be taken into consideration. The tracking or stabilization control of nonholonomic mobile robots and underactuated ships with input saturation is seldom considered in the existing literature. Thus in this thesis the tracking and stabilization control of nonhomonomic mobile robots with parametric uncertainties and external disturbance in the presence of input saturation, and the tracking control of underactuated ships with input saturation, are considered.

For state-feedback control of underactuated mechanical systems the velocities of the mechanical systems must be measured by devices like tachometers. But in practice such devices may not be used either because they are contaminated by noise or they are expensive. Thus we are concerned with the cases where the velocities of the mechanical systems are not needed for the implementation of tracking control. The output feedback tracking control of underactuated mechanical systems like mobile robots and underactuated ships has received attentions from the control community. But it should be pointed out that the output feedback tracking control of nonholonomic mobile robots or underactuated ships with parametric uncertainties has never been considered before. Due to the existence of parametric uncertainties, the output feedback tracking controller for nonholonomic mobile robots and underactuated ships will become much more difficult due to the simultaneous existence of nonholonomic constraints, unknown system parameters and a quadratic term of unmeasurable states in the dynamic system as well as their couplings.

The control of multi-agent systems is receiving a lot of attention from control the community recently. The consensus control of a group of mechanical systems with parametric uncertainties is an interesting topic particularly. An important performance indicator for the consensus problem is the convergence rate. Most of the existing consensus control schemes for multi-agent systems so far achieve asymptotical convergence, namely the convergence rate at best is exponential, which means it needs infinite time for the tracking errors to converge to the origin. Thus designing a controller for each agent such that consensus could be reached within finite time is part of the objective of this thesis. However, achieving adaptive finite-time control is rather challenging. Certain key techniques employed in existing adaptive control literature cannot be applied. To solve adaptive finite-time control problems, some new techniques have to be developed.

Adaptive Control of Underactuated Nonlinear Systems with Input Saturation

2

Adaptive Control of Nonholonomic Mobile Robots with Input Saturation

In this chapter, the stabilization and tracking control of nonholonomic mobile robots with parametric uncertainties and external disturbance in the presence of input saturation is addressed. The torques are designed such that the bounds of the torques are functions of only design parameters and reference trajectory, which are computable in advance. For a given reference trajectory, design parameters are chosen such that the bounds of the torques are within the given saturation limits. To handle the disturbances, the unknown bounds are estimated through adaptive law and the estimates are employed in controller design. The system stability is established.

2.1 Introduction

The two-wheeled mobile robot, a two-input three-output system, is one of the well-known benchmark nonholonomic systems, which leads to the development of various novel controller design schemes. Much effort has been devoted to the stabilization and tracking of this dynamic system, see [7, 14, 31] and the references therein. By means of Brockett's necessary condition [1] for asymptotic stabilization, there exists no continuous time-invariant state-feedback controller that asymptotically stabilizes the mobile robot. Therefore, the past abundant literature aims at developing suitable time-varying controllers for mobile robots. In [32, 6] the stabilization problem is dealt with for a driftless controllable system using bounded state-feedback approach. In [33] the simultaneous stabilization and tracking problem of mobile robots is studied when saturation is imposed on linear and angular velocities. In practice, a mobile robot is ultimately driven by a motor which can only provide a limited amount of torque. Thus the magnitude of the control signal is always constrained. It often severely limits system performance, giving rise to undesirable inaccuracy or leading to instability as in [34, 35]. The development of an adaptive control scheme for a two-wheeled mobile robot with input saturation has been a task of major practical interest as well as theoretical significance. As far as we know, the stabilization and tracking of a nonholonomic mobile robot in the presence of input saturation and external disturbance is rarely addressed in past literature. In [36], a passivity-based, saturated, smooth feedback controller is designed based on the kinematic model of a mobile robot. However, there has been no result addressing the issue of input torque saturation at the dynamic model level for a wheeled mobile robot due to the difficulty of the problem. Also there has been no consideration on external disturbances. The main challenge is how to handle the effects of unknown parameters, the saturation and the external disturbance to achieve internal asymptotic stability and desirable performance.

In this chapter, we consider these effects in solving the stabilization and tracking problem for a wheeled mobile robot. By exploiting the special structure of the mobile robot, a new control scheme is proposed to design controllers and parameter estimators. Through normalizing certain signals appropriately, we are able to make the designed 'saturated' torques bounded by functions of design parameters and reference trajectories only. In this way, we can compute the bounds in advance and thus ensure them to be within the saturation limits by suitably choosing the design parameters. Therefore with these parameters, the control torques are free of saturation and the control law for the 'saturated' torque is the control law we need. To compensate for the disturbances, we estimate their unknown bounds and employ the estimates in controller design. This enables us to achieve perfect tracking of reference trajectories or stabilization to the origin, in addition to the system stability. To illustrate the controller design procedure, simulation studies are conducted for the set-point stabilization and the tracking of a line and a circle in a plane. The obtained numerical results are in line with our theoretical findings.

2.2 System Model and Problem Statement

We consider a wheeled mobile robot shown in Fig.2.1 which is described by the following kinematic and dynamic models

$$\dot{\eta} = J(\eta)\omega \tag{2.1}$$

$$M\dot{\omega} + C(\dot{\eta})\omega + D\omega + \bar{\varepsilon} = sat(\tau) \tag{2.2}$$

where $\eta = (x, y, \phi)$ denotes the position and orientation of the robot, $\omega = (\omega_1, \omega_2)^T$ denotes the angular velocities of the left and right wheels, $\tau = (\tau_1, \tau_2)^T$ represents the control torques applied to the wheels, and M is a symmetric, positive definite inertia matrix, $C(\dot{\eta})$ is the centripetal and coriolis matrix, D denotes the surface friction, and $\bar{\varepsilon}$ represents external time-varying disturbances. The expressions of $J(\eta)$, M, $C(\dot{\eta})$ and D are the same as those in [9], which are also given as below for completeness

$$J(\eta) = \frac{r}{2} \begin{bmatrix} \cos\phi & \cos\phi \\ \sin\phi & \sin\phi \\ b^{-1} & -b^{-1} \end{bmatrix}, M = \begin{bmatrix} m_{11} & m_{12} \\ m_{12} & m_{11} \end{bmatrix}$$

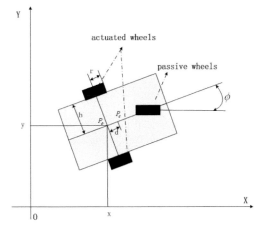

Fig. 2.1. A two-wheeled nonholonomic mobile robot.

$$C(\dot{\eta}) = \begin{bmatrix} 0 & c\dot{\phi} \\ -c\dot{\phi} & 0 \end{bmatrix}, D = \begin{bmatrix} d_{11} & 0 \\ 0 & d_{22} \end{bmatrix}$$

$$m_{11} = 0.25b^{-2}r^2(mb^2 + I) + I_w$$
$$m_{12} = 0.25b^{-2}r^2(mb^2 - I)$$
$$m = m_c + 2m_w$$
$$I = m_c d^2 + 2m_w b^2 + I_c + 2I_m$$
$$c = 0.5b^{-1}r^2 m_c d.$$

In these expressions, b is the half width of the mobile robot and r is the radius of the wheel, d is the distance between the center of the mass, P_c, of the robot and the middle point between the left and right wheels, I_c, I_w and I_m are the moment of inertia of the body about the vertical axis through P_c, the wheel with a motor about the wheel axis, and the wheel with a motor about the diameter, respectively. The positive constants d_{ii}, $i = 1, 2$, are the damping coefficients.

$sat(\tau)$ represents the input torques subject to saturation described by $sat(\tau) = \{sat(\tau_1), sat(\tau_2)\}^T$ and

$$sat(\tau_i) = \begin{cases} sign(\tau_i)u_{Mi} & |\tau_i| \geq u_{Mi} \\ \tau_i & |\tau_i| < u_{Mi} \end{cases}$$

where u_{Mi} is a known saturation bound of τ_i. Let v and w denote the linear and angular velocities of the robot, respectively. Then the relationship between ω_1, ω_2 and v, w is described as follows

$$(v, w)^T = B^{-1}(\omega_1, \omega_2)^T, B = \frac{1}{r} \begin{bmatrix} 1 & b \\ 1 & -b \end{bmatrix}. \tag{2.3}$$

Substituting (2.3) into (2.1), we get

$$\dot{\eta} = L(\eta)\varpi$$
$$\dot{\varpi} = -B^{-1}M^{-1}C(\dot{\eta})B\varpi - B^{-1}M^{-1}DB\varpi + B^{-1}M^{-1}\bar{\varepsilon} + B^{-1}M^{-1}sat(\tau)$$
(2.4)

where

$$L(\eta) = \begin{bmatrix} \cos\phi & 0 \\ \sin\phi & 0 \\ 0 & 1 \end{bmatrix}, \qquad \varpi = [v\ w]^T.$$
(2.5)

Let

$$\bar{B} = \frac{1}{r}\begin{bmatrix} -b & -1 \\ -b & 1 \end{bmatrix}.$$

It is obvious that $C(\dot{\eta})B = \bar{B}C(\dot{\eta})$ and thus we can replace $C(\dot{\eta})B$ in (2.4) by $\bar{B}C(\dot{\eta})$. Multiplying (2.4) by $\bar{B}^{-1}MB$, we get

$$\dot{\eta} = L(\eta)\varpi \tag{2.6}$$
$$R\dot{\varpi} = -\bar{C}(w)\varpi - D_R\varpi + \varepsilon + B_R sat(\tau) \tag{2.7}$$

where $R = -\bar{B}^{-1}MB$, $\bar{C}(w) = -C(w)$, $D_R = -\bar{B}^{-1}DB$, $B_R = -\bar{B}^{-1}$ and R is a positive definite matrix, $\varepsilon = -\bar{B}^{-1}\bar{\varepsilon}$ is a disturbances-like signal. Explicitly, we have

$$R = \begin{bmatrix} (m_{11}+m_{12})/b & 0 \\ 0 & (m_{11}-m_{12})b \end{bmatrix}, \quad D_R = \begin{bmatrix} \frac{d_{11}+d_{22}}{2b} & \frac{d_{11}-d_{22}}{2} \\ \frac{d_{11}-d_{22}}{2} & \frac{b(d_{11}+d_{22})}{2} \end{bmatrix}, \quad B_R = \frac{r}{2b}\begin{bmatrix} 1 & 1 \\ b & -b \end{bmatrix}.$$
(2.8)

Now let $(x_r, y_r, \phi_r)^T$ denote the desired reference position and orientation of the virtual robot the motion of which is described by:

$$\dot{x}_r = v_r \cos\phi_r$$
$$\dot{y}_r = v_r \sin\phi_r$$
$$\dot{\phi}_r = w_r \tag{2.9}$$

where v_r and w_r denote the linear and angular velocities of the the virtual robot. As often done in control of mobile robot [7], we define x_e, y_e and ϕ_e as

$$\begin{bmatrix} x_e \\ y_e \\ \phi_e \end{bmatrix} = \begin{bmatrix} \cos\phi & \sin\phi & 0 \\ -\sin\phi & \cos\phi & 0 \\ 0 & 0 & 1 \end{bmatrix} \begin{bmatrix} x - x_r \\ y - y_r \\ \phi - \phi_r \end{bmatrix}, \tag{2.10}$$

which describe the tracking errors between the actual robot position, direction and those of the virtual robot. It can be directly checked that these tracking errors satisfy the following differential equations:

$$\begin{bmatrix} \dot{x}_e \\ \dot{y}_e \\ \dot{\phi}_e \end{bmatrix} = \begin{bmatrix} wy_e + v - v_r \cos\phi_e \\ -wx_e + v_r \sin\phi_e \\ w - w_r \end{bmatrix}. \tag{2.11}$$

The control objective is to find torque τ applied to the mobile robot, which is subject to saturation, to ensure the desired reference trajectories are tracked in the presence of external disturbances. If v_r and w_r are zero or both converge to zero as $t \to \infty$, the tracking problem becomes set-point stabilization of robots according to (2.10). To solve the control problem, we need the following assumptions:

Assumption 2.1 *The reference linear and angular velocities v_r and w_r and their first-order derivatives are bounded.*

Assumption 2.2 *The unknown parameters of the mobile robot are in known compact sets.*

Assumption 2.3 *The disturbances are bounded, namely, $\varepsilon_1 < E_1$ and $\varepsilon_2 < E_2$ where E_1 and E_2 are positive constants.*

To compensate for the effects of the disturbances, we will obtain an adaptive law to estimate the unknown bounds $E = (E_1, E_2)^T$ for the design of controllers.

2.3 Controller Design

The controller design is divided into two steps. Firstly we develop virtual control laws for v and w based on the kinematic model in such a way that the mobile robot will follow the pre-established reference trajectory. Secondly we derive a design law for $sat(\tau)$ based on the dynamic model of (2.2) so that v and w would converge to the designed virtual controls. Through our design and derivation, suitable normalization is introduced to certain signals. This enables us to make the bound of $sat(\tau)$ as a function of design parameters and reference trajectories only. In this way, we can compute the bound in advance and ensure it within the saturation limit by suitably choosing the design parameters. Therefore with these parameters, $sat(\tau)$ is the same as τ and the derived design law for $sat(\tau)$ is also the control law for τ.

2.3.1 Adaptive control of kinematic model

Introduce a new variable $\bar{\phi}_e = \phi_e + \arcsin(\dfrac{k(t)y_e}{\sqrt{1 + x_e^2 + y_e^2}})$ where

$$k(t) = \lambda_1 v_r + \lambda_2 \sin(t), \tag{2.12}$$

λ_1 and λ_2 are positive constants chosen so that $\lambda_1|v_r|^{\max} + \lambda_2 \leq 0.5$, where $|v_r|^{\max}$ denotes the maximum value of $|v_r|$. This will ensure that $k(t) \leq 0.5$ for all t. Then (2.11) is transformed to

$$\begin{bmatrix} \dot{x}_e \\ \dot{y}_e \\ \dot{\phi}_e \end{bmatrix} = \begin{bmatrix} wy_e + v + f_1 \\ -wx_e - \frac{k(t)v_r y_e}{\Gamma_1} + f_2 \\ w(1 - \frac{k(t)x_e}{\Gamma_1}) + f_3 - v\frac{k(t)x_e y_e}{\Gamma_1^2 \Gamma_2} \end{bmatrix} \tag{2.13}$$

where

$$\Gamma_1 = \sqrt{1 + x_e^2 + y_e^2}, \quad \Gamma_2 = \sqrt{1 + x_e^2 + (1 - k^2(t))y_e^2},$$

$$f_1 = -\frac{v_r \cos\bar{\phi}_e \Gamma_2}{\Gamma_1} - \frac{v_r k(t)y_e \sin\bar{\phi}_e}{\Gamma_1},$$

$$f_2 = -\frac{v_r(\cos\bar{\phi}_e - 1)\Gamma_2}{\Gamma_1} - \frac{v_r k(t)y_e \sin\bar{\phi}_e}{\Gamma_1},$$

$$f_3 = -w_r + \frac{\dot{k}(t)y_e + k(t)f_2 - k(t)(x_e f_1 + y_e f_2)/\Gamma_1^2}{\Gamma_2}. \tag{2.14}$$

Define the virtual control error $\tilde{v}$ and $\tilde{w}$ as

$$\tilde{v} = v - v_c, \quad \tilde{w} = w - w_c \tag{2.15}$$

where v_c and w_c are the virtual controls designed for v and w. We have

$$\begin{bmatrix} \dot{x}_e \\ \dot{y}_e \\ \dot{\phi}_e \end{bmatrix} = \begin{bmatrix} (w_c + \tilde{w})y_e + v_c + \tilde{v} + f_1 \\ -(w_c + \tilde{w})x_e - \frac{k(t)v_r y_e}{\Gamma_1} + f_2 \\ (w_c + \tilde{w})(1 - \frac{k(t)x_e}{\Gamma_1}) + f_3 - (v_c + \tilde{v})\frac{k(t)x_e y_e}{\Gamma_1^2 \Gamma_2} \end{bmatrix}. \tag{2.16}$$

Define a Lyapunov function

$$V_0 = \sqrt{1 + x_e^2 + y_e^2} + \sqrt{1 + \bar{\phi}_e^2} - 2.$$

The time derivative of V_0 is

$$\dot{V}_0 = -\frac{k(t)v_r y_e^2}{\Gamma_1^2} + \frac{x_e}{\Gamma_1}(v_c + f_1) + \frac{y_e f_2}{\Gamma_1} + \frac{\bar{\phi}_e}{\Gamma_3}[w_c(1 - \frac{k(t)x_e}{\Gamma_1}) + f_3$$

$$- v_c \frac{k(t)x_e y_e}{\Gamma_1^2 \Gamma_2}] + \tilde{v}(\frac{x_e}{\Gamma_1} - \frac{k(t)x_e y_e \bar{\phi}_e}{\Gamma_1^2 \Gamma_2 \Gamma_3}) + \tilde{w}\frac{\bar{\phi}_e}{\Gamma_3}(1 - \frac{k(t)x_e}{\Gamma_1})$$

where $\Gamma_3 = \sqrt{1 + \bar{\phi}_e^2}$. The virtual controllers v_c and w_c are chosen as:

$$v_c = -k_1\frac{x_e}{\Gamma_1} - f_1$$

$$w_c = \frac{1}{(1 - \frac{k(t)x_e}{\Gamma_1})}(-k_2\frac{\bar{\phi}_e}{\Gamma_3} - f_3 + v_c\frac{k(t)x_e y_e}{\Gamma_1^2 \Gamma_2}). \tag{2.17}$$

Since we have $k(t) \leq 0.5$ from (2.12) and $\dfrac{x_e}{\Gamma_1} < 1$, so we have $1 - \dfrac{k(t)x_e}{\Gamma_1} > 0.5$. The derivative of $V_0(t)$ is

$$\dot{V}_0 = -k_1 \frac{x_e^2}{\Gamma_1^2} - \frac{k(t)v_r y_e^2}{\Gamma_1^2} - k_2 \frac{\bar{\phi}_e^2}{\Gamma_3^2} + \frac{y_e f_2}{\Gamma_1} + \tilde{v}\left(\frac{x_e}{\Gamma_1} - \frac{k(t)x_e y_e \bar{\phi}_e}{\Gamma_1^2 \Gamma_2}\right)$$
$$+ \tilde{w}\frac{\bar{\phi}_e}{\Gamma_3}\left(1 - \frac{k(t)x_e}{\Gamma_1}\right).$$

Remark 2.1. Γ_1, Γ_2 and Γ_3 are introduced to normalize x_e, y_e and $\bar{\phi}_e$ so that the bounds of virtual control signals based on the kinematic model are only dependent on system parameters and reference signals, and can be pre-determined. This normalization technique will also be applied to the dynamic model when designing control law for $sat(\tau)$.

2.3.2 Adaptive control of dynamic model

Define $\varpi_c = [v_c \ \ w_c]^T$, $\tilde{\varpi} = \varpi - \varpi_c$, and a new Lyapunov function $V_1 = V_0 + \frac{1}{2}\tilde{\varpi}^T R \tilde{\varpi}$, then

$$\dot{V}_1 = \dot{V}_0 + \tilde{\varpi}^T R \dot{\tilde{\varpi}} = \dot{V}_0 + \tilde{\varpi}^T R(\dot{\varpi} - \dot{\varpi}_c)$$
$$= \dot{V}_0 + \tilde{\varpi}^T[-\bar{C}(w)\varpi - D_R\varpi + \varepsilon + B_R sat(\tau) - R\dot{\varpi}_c]$$
$$= \dot{V}_0 + \tilde{\varpi}^T[-(\bar{C}(w) + D_R))\tilde{\varpi} + \varepsilon + B_R sat(\tau) - Y_c P] \qquad (2.18)$$

where $Y_c P = R\dot{\varpi}_c + C(w)\varpi_c + D_R\varpi_c$ and

$$Y_c = \begin{bmatrix} \dot{v}_c & 0 & w_c w & v_c & w_c & 0 \\ 0 & \dot{w}_c & -v_c w & 0 & v_c & w_c \end{bmatrix}, \qquad (2.19)$$

$$P = \begin{bmatrix} \frac{m_{11}+m_{22}}{b} & (m_{11}-m_{22})b & c & \frac{d_{11}+d_{22}}{2b} & \frac{d_{11}-d_{22}}{2} & \frac{b(d_{11}+d_{22})}{2} \end{bmatrix}^T. \qquad (2.20)$$

Remark 2.2. Unlike the typical way of designing virtual controllers for (2.11) as in [9, 7], we use v and w as the virtual controllers. By doing this, the tracking error dynamic system in (2.13) and the virtual controllers v_c and w_c in (2.17) do not contain any unknown system parameters. So when taking the time-derivatives of v_c and w_c in the matrix Y_c in (2.19) to get the torque τ_i, no unknown parameter will appear. This will reduce the number of unknown parameters in (2.20) from 10 in [9] to 6 and thus greatly simplify the matrix Y_c. Therefore parameter estimation is simplified compared to [9].

Denote $\alpha := \dfrac{r}{2b}$, $\beta := \dfrac{r}{2}$ and $m = \dfrac{1}{\alpha}$, $n = \dfrac{1}{\beta}$. Let $\hat{m}$ and $\hat{n}$ be the estimates of m and n respectively, and define $\tilde{m} = m - \hat{m}$, $\tilde{n} = n - \hat{n}$, $u_1 = \dfrac{sat(\tau_1) + sat(\tau_2)}{\hat{m}}$, $u_2 = \dfrac{sat(\tau_1) - sat(\tau_2)}{\hat{n}}$, $U = [u_1 \ u_2]^T$. Then we get

$$B_R sat(\tau) = \begin{bmatrix} \alpha & \alpha \\ \beta & -\beta \end{bmatrix} \begin{bmatrix} sat(\tau_1) \\ sat(\tau_2) \end{bmatrix} = \begin{bmatrix} \alpha \hat{m} u_1 \\ \beta \hat{n} u_2 \end{bmatrix}.$$

As $\alpha \hat{m} = \alpha(m - \tilde{m}) = 1 - \alpha \tilde{m}$ and also $\beta \hat{n} = 1 - \beta \tilde{n}$,

$$B_R sat(\tau) = \begin{bmatrix} u_1 \\ u_2 \end{bmatrix} - \begin{bmatrix} \alpha \tilde{m} u_1 \\ \beta \tilde{n} u_2 \end{bmatrix} = U - \begin{bmatrix} \alpha \tilde{m} u_1 \\ \beta \tilde{n} u_2 \end{bmatrix}. \tag{2.21}$$

Let $\Omega_1 = (\dfrac{x_e}{\Gamma_1} - \dfrac{k(t)x_e y_e \bar{\phi}_e}{\Gamma_1^2 \Gamma_2 \Gamma_3})$, $\Omega_2 = \dfrac{\bar{\phi}_e}{\Gamma_3}(1 - \dfrac{k(t)x_e}{\Gamma_1})$, and $\Omega = [\Omega_1 \;\; \Omega_2]^T$. We now define another new Lyapunov function

$$V_2 = V_1 + \frac{1}{2}\tilde{P}^T T \tilde{P} + \frac{1}{2}\tilde{E}^T K \tilde{E} + \frac{1}{2}\alpha \tilde{m}^2 + \frac{1}{2}\beta \tilde{n}^2$$

where $\tilde{P} = P - \hat{P}$ and $\hat{P}$ is the estimate of P, $T = aI$ with a being a positive constant; $\tilde{E} = E - \hat{E}$, $\hat{E}$ is the estimates of the bounds of unknown disturbances, K is a given positive definite matrix. Then

$$\dot{V}_2 = \dot{V}_0 + \tilde{\varpi}^T[-(\bar{C}(w) + D_R))\tilde{\varpi} + \varepsilon + B_R sat(\tau) - Y_c P]$$
$$+ \tilde{P}^T T \dot{\tilde{P}} + \tilde{E}^T K \dot{\tilde{E}} + \alpha \tilde{m}\dot{\tilde{m}} + \beta \tilde{n}\dot{\tilde{n}}$$
$$= -k_1 \frac{x_e^2}{\Gamma_1^2} - \frac{k(t)v_r y_e^2}{\Gamma_1^2} - k_2 \frac{\bar{\phi}_e^2}{\Gamma_3^2} + \frac{y_e f_2}{\Gamma_1} + \alpha \tilde{m}\dot{\tilde{m}} + \beta \tilde{n}\dot{\tilde{n}}$$
$$+ \tilde{\varpi}^T[-(\bar{C}(w) + D_R))\tilde{\varpi} + \varepsilon + B_R sat(\tau) - Y_c P + \Omega] + \tilde{P}^T T \dot{\tilde{P}} - \tilde{E}^T K \dot{\tilde{E}}$$

$$= -k_1 \frac{x_e^2}{\Gamma_1^2} - \frac{k(t)v_r y_e^2}{\Gamma_1^2} - k_2 \frac{\bar{\phi}_e^2}{\Gamma_3^2} + \frac{y_e f_2}{\Gamma_1} - \alpha \tilde{m}(\dot{\tilde{m}} + u_1 \tilde{v}) - \beta \tilde{n}(\dot{\tilde{n}} + u_2 \tilde{w})$$
$$+ \tilde{P}^T T \dot{\tilde{P}} - \tilde{E}^T K \dot{\tilde{E}}$$
$$+ \tilde{\varpi}^T[-(\bar{C}(w) + D_R))\tilde{\varpi} + \varepsilon + U - Y_c P + \Omega]. \tag{2.22}$$

By choosing U as

$$U = Y_c \hat{P} - \Omega - sgn(\tilde{\varpi})^T * \hat{E} \tag{2.23}$$

where $sgn(\tilde{\varpi})^T * \hat{E} = (sgn(\tilde{\varpi}_1)\hat{E}_1, \;\; sgn(\tilde{\varpi}_2)\hat{E}_2)^T$, we get

$$\dot{V}_2 \leq -k_1 \frac{x_e^2}{\Gamma_1^2} - \frac{k(t)v_r y_e^2}{\Gamma_1^2} - k_2 \frac{\bar{\phi}_e^2}{\Gamma_3^2} + \frac{y_e f_2}{\Gamma_1} - \tilde{\varpi}^T D_R \tilde{\varpi}$$
$$- \alpha \tilde{m}(\dot{\tilde{m}} + u_1 \tilde{v}) - \beta \tilde{n}(\dot{\tilde{n}} + u_2 \tilde{w}) - \tilde{P}^T T (\dot{\hat{P}} + T^{-1}Y_c^T \tilde{\varpi})$$
$$- \tilde{E}^T K (\dot{\hat{E}} - K^{-1}|\tilde{\varpi}|).$$

The torque τ is designed as

$$sat(\tau) = \begin{bmatrix} 1 & 1 \\ 1 & -1 \end{bmatrix}^{-1} \begin{bmatrix} \hat{m} & 0 \\ 0 & \hat{n} \end{bmatrix} U. \tag{2.24}$$

The initial values of estimates are set to be in the given compact sets. By using parameter projection, the estimates to be updated are confined in the known sets. To illustrate this, we use the update laws for $\hat{m}$ and $\hat{n}$ as examples. From Assumption 2.2, there exist $m_{\max}, m_{\min}, n_{\max}, n_{\min}$ and a small positive constant σ so that $m_{\min} + \sigma < m < m_{\max} - \sigma$ and $n_{\min} + \sigma < n < n_{\max} - \sigma$. The update laws for $\hat{m}$ and $\hat{n}$ are designed as follows:

$$\dot{\hat{m}} = \begin{cases} -u_1\tilde{v} & m_{\min} + \sigma < \hat{m} < m_{\max} - \sigma \\ -u_1\tilde{v} + \frac{1}{2}[1 + (u_1\tilde{v})^2] & \hat{m} \leq m_{\min} + \sigma \\ -u_1\tilde{v} - \frac{1}{2}[1 + (u_1\tilde{v})^2] & \hat{m} \geq m_{\max} - \sigma \end{cases}$$

$$\dot{\hat{n}} = \begin{cases} -u_2\tilde{w} & n_{\min} + \sigma < \hat{n} < n_{\max} - \sigma \\ -u_2\tilde{w} + \frac{1}{2}[1 + (u_2\tilde{w})^2] & \hat{n} \leq n_{\min} + \sigma \\ -u_2\tilde{w} - \frac{1}{2}[1 + (u_2\tilde{w})^2] & \hat{n} \geq n_{\max} - \sigma. \end{cases} \tag{2.25}$$

We choose the initial value for $\hat{m}$ and $\hat{n}$ so that $m_{\min} < \hat{m}(0) < m_{\max}$ and $n_{\min} < \hat{n}(0) < n_{\max}$. When $\hat{m} \leq m_{\min} + \sigma$, we have $\dot{\hat{m}} \geq 0$, and this law will prevent $\hat{m}$ from getting smaller than $m_{\min}$. Also we get $\tilde{m} = m - \hat{m} > 0$ and $\dot{\hat{m}} + u_1\tilde{v} > 0$. Therefore we have $-\alpha\tilde{m}(\dot{\hat{m}} + u_1\tilde{v}) < 0$. Similarly, when $\hat{m} \geq m_{\max} - \sigma$, we have $\dot{\hat{m}} \leq 0$, and this law will prevent $\hat{m}$ from getting greater than $m_{\max}$. Also we get $\tilde{m} = m - \hat{m} < 0$ and $\dot{\hat{m}} + u_1\tilde{v} < 0$, and thus $-\alpha\tilde{m}(\dot{\hat{m}} + u_1\tilde{v}) < 0$. For other parameters the same parameter projection rule is adopted. From the above analysis, we have

$$\dot{V}_2 \leq -k_1 \frac{x_e^2}{\Gamma_1^2} - \frac{k(t)v_r y_e^2}{\Gamma_1^2} - k_2 \frac{\bar{\phi}_e^2}{\Gamma_3^2} + \frac{y_e f_2}{\Gamma_1}$$
$$- \tilde{\omega}^T D_R \tilde{\omega} - \tilde{P}^T T(\dot{\hat{P}} + T^{-1} Y_c^T \tilde{\omega}) - \tilde{E}^T K(\dot{\hat{E}} - K^{-1}|\tilde{\omega}|).$$

From Assumption 2.2, we have P in a known set which satisfies $P(i)_{\min} + \sigma < P(i) < P(i)_{\max} - \sigma$ and $E(j) < E_{\max}(j) - \sigma$ for $i = 1, 2, 3, 4, 5$ and $j = 1, 2$. The parameters update laws for $\hat{P}$ and $\hat{E}$ are then chosen as

$$\dot{\hat{P}}(i) = \begin{cases} \Psi(i) & P(i)_{\min} + \sigma < \hat{P}(i) < P(i)_{\max} - \sigma \\ \Psi(i) + \frac{1}{2}(1 + \Psi(i)^2) & \hat{P}(i) \leq P(i)_{\min} + \sigma \\ \Psi(i) - \frac{1}{2}(1 + \Psi(i)^2) & \hat{P}(i) \geq P(i)_{\max} - \sigma \end{cases}$$

$$\dot{\hat{E}}(j) = \begin{cases} \Xi(j) & \sigma \leq \hat{E}(j) < E_{\max}(j) - \sigma \\ \Xi(j) - \frac{1}{2}(1 + \Xi(j)^2) & \hat{E}(j) \geq E_{\max}(j) - \sigma \\ \Xi(j) + \frac{1}{2}(1 + \Xi(j)^2) & \hat{E}(j) < \sigma \end{cases} \tag{2.26}$$

where $\Psi = -T^{-1}Y_c^T\tilde{\omega}$ and $\Xi = K^{-1}|\tilde{\omega}|$. Following the same analysis as (2.25), these parameters update laws guarantee that $P(i)_{\min} < \hat{P}(i) <$

$P(i)_{\max}$, $E(j) < E_{\max}(j)$ and $-\tilde{P}^T T(\dot{\hat{P}} + T^{-1}Y_c^T\tilde{\varpi}) - \tilde{E}^T K(\dot{\hat{E}} - K^{-1}|\tilde{\varpi}|) \leq 0$. Thus we have

$$\dot{V}_2 \leq -k_1\frac{x_e^2}{\Gamma_1^2} - \frac{k(t)v_r y_e^2}{\Gamma_1^2} - k_2\frac{\bar{\phi}_e^2}{\Gamma_3^2} + \frac{y_e f_2}{\Gamma_1} - \tilde{\varpi}^T D_R \tilde{\varpi}. \qquad (2.27)$$

Proposition 2.3. Under the controllers (2.17) and (2.24) with parameter update laws in (2.25) and (2.26), there exist appropriate design constants k_1, k_2, λ_1, λ_2, δ_1 and δ_2 such that system (2.16) is uniformly stable and the trajectory of virtual robot (2.9) is asymptotically tracked or the system is stabilized to the origin..

Proof. It is proved in the Appendix that positive constants $\delta_1 \in (0, 1)$ and $\delta_2 > 0$ can be chosen such that

$$\frac{y_e f_2}{\Gamma_1} \leq \frac{\delta_1 v_r{}^2 y_e{}^2}{\Gamma_1{}^2} + \delta_2 \frac{\bar{\phi}_e^2}{\Gamma_3^2}. \qquad (2.28)$$

Substituting (2.28) into (2.27) yields

$$\dot{V}_2 \leq -k_1\frac{x_e^2}{\Gamma_1^2} - \frac{(\lambda_1 - \delta_1)v_r^2 y_e^2}{\Gamma_1^2} - (k_2 - \delta_2)\frac{\bar{\phi}_e^2}{\Gamma_3^2} - \tilde{\varpi}^T D_R \tilde{\varpi}$$
$$- \frac{\lambda_2 \sin(t)v_r y_e^2}{\Gamma_1^2}. \qquad (2.29)$$

On the other hand, based on the condition that $|k(t)| < 0.5$, we choose positive constants λ_1 and λ_2 to satisfy

$$\lambda_1 v_r{}^{\max} + \lambda_2 < 0.5 \qquad (2.30)$$

where $v_r{}^{\max}$ denotes the maximum value of $|v_r|$. From (2.29), the choices of λ_1, λ_2 and k_2 also meet the following condition

$$\lambda_1 > \delta_1, \quad k_2 > \delta_2, \quad v_r{}^2(\lambda_1 - \delta_1) - |\lambda_2 v_r| \geq 0.$$

Clearly the above choices of parameters are possible, and we choose $k_1 > 0$. Let $c_1 = k_2 - \delta_2$ and $c_2 = v_r{}^2(\lambda_1 - \delta_1) - |\lambda_2 v_r|$, then we get

$$\dot{V}_2 \leq -\frac{c_2 y_e{}^2}{\Gamma_1{}^2} - \frac{k_1 x_e{}^2}{\Gamma_1{}^2} - \frac{c_1 \bar{\phi}_e{}^2}{\Gamma_3{}^2} - \tilde{\varpi}^T D_R \tilde{\varpi}. \qquad (2.31)$$

For stability analysis we consider the stabilization and tracking separately. For tracking, we have $\lim_{t\to\infty} |v_r| \neq 0$, thus we get $\lim_{t\to\infty} c_2 > 0$. Then from (2.31) we get $\lim_{t\to\infty} |x_e| + |y_e| + |\bar{\phi}_e| \to 0$.

For stabilization, we have $\lim\limits_{t\to\infty} |v_r| + |\dot{v}_r| + |w_r| + |\dot{w}_r| = 0$. So from (2.31) we get

$$\dot{V}_2 \le -\frac{k_1 x_e{}^2}{\Gamma_1{}^2} - c_1 \frac{\bar{\phi}_e{}^2}{\Gamma_3{}^2} - \tilde{\varpi}^T D_R \tilde{\varpi} \le 0 \tag{2.32}$$

From (2.32), $(x_e, y_e, \bar{\phi}_e)$ is bounded and $\lim\limits_{t\to\infty} |x_e| + |\bar{\phi}_e| + |\tilde{v}| + |\tilde{w}| \to 0$. It is left to prove that $\lim\limits_{t\to\infty} |y_e| \to 0$. From (2.16) and (2.17) we get

$$\dot{x}_e = -\frac{k_1 x_e}{\Gamma_1} + \Sigma(t) \tag{2.33}$$

where

$$\Sigma(t) = \frac{-y_e}{(1 - \frac{k(t)x_e}{\Gamma_1})} \left(k_2 \frac{\bar{\phi}_e}{\Gamma_3} + f_3 + (k_1 \frac{x_e}{\Gamma_1} + f_1) \frac{k(t) x_e y_e}{\Gamma_1{}^2 \Gamma_2} \right) + \tilde{w} y_e + \tilde{v}.$$

Based on (2.7) and (2.11) it is easy to check that $\ddot{x}_e$ is bounded. Since $\lim\limits_{t\to\infty} x_e = 0$, by using Barbalat's Lemma we get $\lim\limits_{t\to\infty} \dot{x}_e = 0$, which means $\lim\limits_{t\to\infty} \Sigma(t) - \frac{k_1 x_e}{\Gamma_1} = 0$. Since $\lim\limits_{t\to\infty} \tilde{v} = \lim\limits_{t\to\infty} \tilde{w} = 0$, then from (2.14) and (2.33)

$$\lim\limits_{t\to\infty} \frac{k(t) y_e{}^2}{\Gamma_2} = 0. \tag{2.34}$$

If λ_1 and λ_2 are set so that $\lim\limits_{t\to\infty} \dot{k}(t) \ne 0$, (2.34) implies $\lim\limits_{t\to\infty} y_e = 0$. As $k(t) = \lambda_1 v_r + \lambda_2 \sin(t)$ and λ_2 is free of choice, this condition can be satisfied if $\dot{v}_r$ is not linearly related to function $\sin(t)$. Otherwise, λ_1 and λ_2 should be carefully chosen to prevent $\lambda_1 \dot{v}_r$ and $\lambda_2 \cos(t)$ from canceling each other.

Finally from the definition of $\bar{\phi}_e$ we know that $\lim\limits_{t\to\infty} y_e + |\bar{\phi}_e| \to 0$ guarantees $\lim\limits_{t\to\infty} |\phi_e| \to 0$. Therefore the trajectories are asymptotically tracked or the robot is stabilized to the origin. $\qquad\square$

Note that $\frac{1}{2} r_{\min} \varpi^T \varpi \le \frac{1}{2} \varpi^T R \varpi \le \frac{1}{2} r_{\max} \varpi^T \varpi$ where $r_{\max} = \max(R_{11}, R_{22})$, $r_{\min} = \min(R_{11}, R_{22})$. Taking the derivative of $V_\varpi = \frac{1}{2}\varpi^T R \varpi$, we get

$$\begin{aligned}
\dot{V}_\varpi &= \varpi^T R \dot{\varpi} = \varpi^T(-(\bar{C}(\dot{\eta}) + D_R)\varpi + \varepsilon + B_R sat(\tau)) \\
&= -\varpi^T D_R \varpi + \varpi^T \bar{s}(\tau) \\
&\le -d_{\min}\varpi^T \varpi + 0.5\zeta \varpi^T \varpi + \frac{0.5}{\zeta}\bar{s}(\tau)^T \bar{s}(\tau) \\
&= -d_\zeta \varpi^T \varpi + \frac{0.5}{\zeta}\bar{s}(\tau)^T \bar{s}(\tau) \\
&\le -\frac{d_\zeta}{r_{\max}} V_\varpi + \frac{0.5}{\zeta}\bar{s}(\tau)^T \bar{s}(\tau)
\end{aligned} \tag{2.35}$$

where $d_{\min}$ is the smallest eigenvalue of D_R, $d_\zeta = d_{\min} - 0.5\zeta$ and $\bar{s}(\tau) = B_R sat(\tau) + \varepsilon$ with ζ being any positive constant that satisfies $d_{\min} > 0.5\zeta$. Then we have

$$V_{\varpi} \le e^{-\frac{d_{\zeta}}{r_{\max}}t}V_{\varpi}(0) + \frac{\overline{s}(\tau)^T\overline{s}(\tau)r_{\max}}{2\zeta d_{\zeta}}. \tag{2.36}$$

So from (2.35) and (2.36) we obtain

$$|\varpi_i| \le \sqrt{\frac{2}{r_{\min}}\left(e^{-\frac{d_{\zeta}}{r_{\max}}t}V_{\varpi}(0) + \frac{\overline{s}(\tau)^T\overline{s}(\tau)r_{\max}}{2\zeta d_{\zeta}}\right)} \tag{2.37}$$

for $i = 1, 2$.

From (2.14) and (2.17) we know that $\dot{\varpi}_c$ and ϖ_c are bounded. From (2.27) $\hat{P}$ is bounded, and from (2.37) w is bounded. Thus the torque designed for $sat(\tau)$ is bounded and its bound is given in the following proposition.

Proposition 2.4. *For the designed torque $sat(\tau_i)$, $i = 1, 2$, there exists a bound $\tau_i{}^{\max}$ such that $|sat(\tau_i)| < \tau_i{}^{\max}$ and $\tau_i{}^{\max}$ is solely a function of designed parameters, reference trajectories and the bounds of the known compact sets in Assumption 2.2, namely*

$$\tau_i{}^{\max} = \mathcal{F}(v_r{}^{\max}, \ w_r{}^{\max}, \ \dot{v}_r{}^{\max}, \ \dot{w}_r^{\max}, \ \lambda_1, \lambda_2, \ k_1, \ k_2,$$
$$\hat{m}_{\max}, \hat{n}_{\max}, \ \hat{P}_{\max}, \ \hat{E}_{\max}, \ \varpi_{\max}), i = 1, 2 \tag{2.38}$$

where

$$\mathcal{F}(\cdot) = (\hat{m}_{\max}\hat{P}_{3\max}w_{\max} + \hat{P}_{5\max} + \hat{n}_{\max}\hat{P}_{6\max})(k_2 + 1.15(0.2\dot{v}_r^{\max} + 0.1)$$
$$+ 2.7v_r^{\max} + w_r^{\max} + 0.5(k_1 + 1.11v_r^{\max}))$$
$$+ (\hat{m}_{\max}\hat{P}_{4\max} + \hat{P}_{3\max}\hat{n}_{\max}w_{\max})(0.5k_1 + 0.56v_r^{\max})$$
$$+ (\hat{m}_{\max}\hat{P}_{1\max})((1.62k_1 + \lambda_1)v_r^{\max} + 1.52v_r^{2\,\max} + 12.8\dot{v}_r^{\max})$$
$$+ \hat{n}_{\max}\hat{P}_{2\max}(0.85w_r^{2\,\max} + 1.02v_r^{2\,\max} + 2.5\dot{w}_r^{\max}$$
$$+ 4.63v_r^{\max}w_r^{\max} + \lambda_1\dot{v}_r^{\max} + \lambda_2$$
$$+ 2.1\lambda_1v_r^{\max}\dot{v}_r^{\max} + 2.1v_r^{\max}\lambda_2)$$
$$+ 0.5\hat{m}_{\max}(1 + \hat{E}_{1\max} + 0.5\hat{n}_{\max}(1.5 + \hat{E}_{2\max}))$$

and $v_r{}^{\max}$, $w_r{}^{\max}$, $\dot{v}_r^{\max}$, $\dot{w}_r^{\max}$, $\hat{P}_{i\max}$, $1 \le i \le 6$ denote the maximum values of v_r, w_r, $\dot{v}_r$, $\dot{w}_r$ and P_i respectively.

Proof. From (2.23) and (2.24) τ_i is composed of the estimated parameters, ϖ_c and $\dot{\varpi}_c$. From (2.12), (2.13), (2.14) and (2.17), ϖ_c and $\dot{\varpi}_c$ only depend on variables $\dfrac{d}{dt}\dfrac{x_e}{\Gamma_1}$, $\dfrac{d}{dt}\dfrac{y_e}{\Gamma_1}$, $\dfrac{d}{dt}\dfrac{\phi_e}{\Gamma_3}$, $\dfrac{d}{dt}\dfrac{x_e}{\Gamma_2}$, $\dfrac{d}{dt}\dfrac{y_e}{\Gamma_2}$, $\dfrac{d}{dt}\dfrac{1}{\Gamma_2}$, $\dfrac{d}{dt}\dfrac{\Gamma_2}{\Gamma_1}$, v_r, w_r, $\dot{v}_r$, $\dot{w}_r$, λ_1, λ_2, k_1, k_2, $\dfrac{x_e}{\Gamma_1}$, $\dfrac{y_e}{\Gamma_1}$, $\dfrac{\phi_e}{\Gamma_3}$, $\dfrac{1}{\Gamma_2}$, $\dfrac{\Gamma_2}{\Gamma_1}$, $\dfrac{x_e}{\Gamma_2}$, $\dfrac{y_e}{\Gamma_2}$, $\cos(\bar{\phi}_e)$, $\sin(\bar{\phi}_e)$, $k(t)$, $\dot{k}(t)$, $\ddot{k}(t)$ and ϖ. These signals are bounded and their bounds are computable. For example, note that $\dfrac{d}{dt}\dfrac{x_e}{\Gamma_1} = \dfrac{\dot{x}_e}{\Gamma_1} - \dfrac{x_e}{\Gamma_1}(\dfrac{x_e}{\Gamma_1}\dfrac{\dot{x}_e}{\Gamma_1} + \dfrac{y_e}{\Gamma_1}\dfrac{\dot{y}_e}{\Gamma_1})$. From (2.13) we know

the bounds of $\dfrac{\dot{x}_e}{\Gamma_1}$ and $\dfrac{\dot{y}_e}{\Gamma_1}$ are computable. Thus the bound of $\dfrac{d}{dt}\dfrac{x_e}{\Gamma_1}$ is also computable. It can be similarly checked that the bounds of other ones are also computable. Since $|\dfrac{x_e}{\Gamma_1}| < 1$, $|\dfrac{y_e}{\Gamma_1}| < 1$, $|\dfrac{\bar{\phi}_e}{\Gamma_3}| < 1$, $|\dfrac{1}{\Gamma_2}| < 1$, $|\dfrac{\Gamma_2}{\Gamma_1}| < 1$, $|\dfrac{x_e}{\Gamma_2}| < 1$, $\cos(\bar{\phi}_e)| \leq 1$, $\sin(\bar{\phi}_e)| \leq 1$ and $|\dfrac{y_e}{\Gamma_2}| < 1.15$, from (2.24) we know (2.38) holds. Replacing these variables with their maximum values and after a lengthy calculation, we get the expression of $\mathcal{F}(\cdot)$.   □

From $\mathcal{F}(\cdot)$ we can calculate $\tau_i{}^{\max}$ and ensure $\tau_i{}^{\max} \leq u_{Mi}$ by choosing suitable parameters and reference signals. If $\tau_i{}^{\max} \leq u_{Mi}$, we have $sat(\tau_i) = \tau_i$, $i = 1, 2$. In this case, the design law in (2.24) is our control law for τ, namely

$$\tau = \begin{bmatrix} 1 & 1 \\ 1 & -1 \end{bmatrix}^{-1} \begin{bmatrix} \hat{m} & 0 \\ 0 & \hat{n} \end{bmatrix} U. \tag{2.39}$$

Remark 2.5. For a general class of system, it is difficult to establish a bound for the control signal as a function of design parameters. The result of Proposition 2.4 is achieved mainly due to exploiting the special structure of the mobile robot with three ideas. Firstly, $sat(\tau_i)$, $i = 1, 2$ is so designed that x_e, y_e and $\bar{\phi}_e$ appearing in the design law are normalized. Secondly, the proposed parameters update law guarantees the unknown parameters are in known compact sets and thus bounded. Thirdly, by using v and w as the virtual controllers, designing parameter update laws and the control law for τ is greatly simplified, which enables us to compute the bound of τ possible.

Remark 2.6. The actual form of $\mathcal{F}(\cdot)$ in (2.38) depends on the given system and reference trajectories. It is noted that $\tau_i^{\max}$ can be divided into two parts: one totally independent of all design parameters λ_1, λ_2, k_1 and k_2 mentioned in Proposition 2.3, and the other only having these design parameters. If the former one is smaller than the saturation bound, then there is a margin so that we can find such design parameters to make $\tau_i^{\max} < u_{Mi}$. How to obtain $\mathcal{F}(\cdot)$ and calculate $\tau_i{}^{\max}$ will be illustrated in simulation studies.

Theorem 2.7. *Consider the nonholonomic system (11.1) and (11.2) with input saturation and disturbance satisfying Assumptions 2.1 - 2.3. With the application of controller (2.24) and parameters update law (2.25) and (2.26), the set-point stabilization and the asymptotical tracking for a reference are achieved simultaneously, and the following statements hold:*
(1) The steady state virtual control error satisfies

$$\lim_{t\to\infty} \tilde{\varpi}^T\tilde{\varpi} = \lim_{t\to\infty} (\varpi - \varpi_c)^T(\varpi - \varpi_c) = 0 \tag{2.40}$$

(2) The bound of virtual control error at any time is given by

$$|\varpi_i - \varpi_{ic}| \le \sqrt{\frac{2}{r_{\min}}} \times$$

$$\sqrt{e^{-\frac{2d_{\min}}{r_{\max}}t}V_3(0) + \frac{P_m{}^T T P_m + E_m{}^T K E_m + \alpha m_m^2 + \beta n_m^2}{2} + \frac{r_{\max}(\varpi_{\max} + \varpi_c{}^m)^T \Omega_m}{2d_{\min}}}$$

(2.41)

for $i = 1, 2$, where $m_m = m_{\max} - m_{\min}$, $n_m = n_{\max} - n_{\min}$, $P_m = P_{\max} - P_{\min}$ and $E_m = E_{\max} - E_{\min}$, $\varpi_{\max}$ is defined in (2.37), $\varpi_c{}^m$ and Ω_m are the maximum values of ϖ_c and Ω respectively, and V_3 is defined in (2.42).

Proof:
(1). Based on *Propositions* 2.3 and 2.4, we can ensure that the set-point stabilization and the asymptotical tracking for a reference are achieved simultaneously. From (2.29) we know that V_2 is non-increasing. Hence $\tilde{\varpi}$, $\hat{P}$ and $\hat{E}$ are bounded. By applying the LaSalle-Yoshizawa theorem to (2.29), we further know that $\tilde{\varpi}^T \tilde{\varpi} \to 0$ as $t \to \infty$, which implies (2.40) holds.
(2). Define a new Lyapunov function

$$V_3 = \frac{1}{2}\tilde{\varpi}^T R \tilde{\varpi} + \frac{1}{2}\tilde{P}^T T \tilde{P} + \frac{1}{2}\tilde{E}^T K \tilde{E} + \frac{1}{2}\alpha\tilde{m}^2 + \frac{1}{2}\beta\tilde{n}^2. \qquad (2.42)$$

We get $\dot{V}_3 \le -\tilde{\varpi}^T D_R \tilde{\varpi} - \tilde{\varpi}^T \Omega$. Since $-\tilde{\varpi}^T D_R \tilde{\varpi} \le -d_{\min}\varpi^T \varpi \le -\frac{2d_{\min}}{r_{\max}}\frac{1}{2}\tilde{\varpi}^T R \tilde{\varpi}$, we have

$$\dot{V}_3 \le -\frac{2d_{\min}}{r_{\max}}V_3 + \frac{d_{\min}}{r_{\max}}(\tilde{P}^T T \tilde{P} + \tilde{E}^T K \tilde{E} + \alpha\tilde{m}^2 + \beta\tilde{n}^2) + (\varpi_{\max} + \varpi_c{}^m)^T \Omega_m.$$

Thus we obtain

$$V_3(t) \le e^{-\frac{2d_{\min}}{r_{\max}}t}V_3(0) + \frac{P_m{}^T T P_m + E_m{}^T K E_m + \alpha m_m^2 + \beta n_m^2}{2}$$
$$+ \frac{r_{\max}(\varpi_{\max} + \varpi_c{}^m)^T \Omega_m}{2d_{\min}}. \qquad (2.43)$$

Also as we have $\frac{1}{2}\tilde{\varpi}^T R \tilde{\varpi} \ge \frac{r_{\min}}{2}\tilde{\varpi}^T \tilde{\varpi}$. This gives that

$$\tilde{\varpi}^T \tilde{\varpi} \le \frac{2}{r_{\min}}\left(e^{-\frac{2d_{\min}}{r_{\max}}t}V_3(0) + \frac{P_m{}^T T P_m + E_m{}^T K E_m + \alpha m_m^2 + \beta n_m^2}{2}\right.$$
$$+ \left.\frac{r_{\max}(\varpi_{\max} + \varpi_c{}^m)^T \Omega_m}{2d_{\min}}\right). \qquad (2.44)$$

Thus it can be obtained that (2.41) holds. $\qquad \square$

2.4 Simulation Results

In this section, we illustrate the design procedure and how to compute the bound of the control torque based on design parameters and reference trajectories. Our simulation tool is MATLAB.

The physical parameters of robots are set as: $b = 0.75$, $d = 0.3$, $r = 0.5$, $m_c = 10$, $m_w = 1$, $I_c = 5.6$, $I_w = 0.005$, $I_m = 0.0025$, $d_{11} = d_{22} = 1$. The unknown disturbance ε is white noise satisfying $0 \le |\varepsilon_i| \le 2$ for $i = 1, 2$. The initial estimates for the unknown parameters are set to be 75% of their true values.

Suppose the torque saturation bound u_{Mi} is 100, $i = 1, 2$. Based on Proposition 2.4, we need to choose suitable parameters in order to perfectly track a given reference and guarantee the designed torque strictly within the saturation bound. For simplicity, we only leave parameter k_1 as a free parameter to achieve this. The other parameters are chosen as $k_2 = 3$, $\lambda_1 = 0.2$, $\lambda_2 = 0.1$, $T = 0.06I$, $K = 0.03I$. The maximum and minimum values of parameters r, b, m_{11}, m_{12}, c, d_{11}, d_{22} are given as $\{(0.5, 1.5), (1, 2), (0.1, 0.5), (-0.1, 0.05), (0.05, 0.2), (0.2, 1.5), (0.2, 1.5)\}$ respectively. Set the initial velocities $\{\varpi_1, \varpi_2\} = \{0, 0\}$, then we have $V_w(0) = 0$. From (2.37) we know

$$|\varpi_{1,2}| \le \sqrt{\frac{2}{r_{\min}} \frac{\overline{s}(\tau)^T \overline{s}(\tau) r_{\max}}{2\zeta d_\zeta}} \le \sqrt{\frac{\overline{s}(\tau)^T \overline{s}(\tau)}{2\zeta(5 - 0.5\zeta)}}$$

Choose $\zeta = 5$, we get $|\varpi_{1,2}| \le 28.31$, which means $|w| \le 28.31$. Note that $f_1 < 1.11v_r$, $f_2 < 2.1v_r$, $f_3 < 1.15(0.2\dot{v}_r + 0.1) + 2.7v_r + w_r$, $v_c < k_1 + f_1$ and $w_c < 2(3 + f_3 + 0.5v_c)$. Replacing $\hat{m}$, $\hat{n}$, m_{11}, m_{12}, c, d_{11}, d_{22}, w by their maximum values specified above, after a lengthy but simple calculation, we get polynomial $\mathcal{F}(\cdot)$ for computing $\tau_i^{\max}$

$$\tau_i^{\max} = 4.13v_r^2 + (30.34 + 0.26k_1)v_r + 10.35 + 0.75k_1$$
$$+ 2.04w_r^2 + 11.12v_r w_r + 15.2\dot{v}_r + 6\dot{w}_r$$
$$= \mathcal{F}_1(v_r, w_r, \dot{v}_r, \dot{w}_r) + \mathcal{F}_2(k_1) \qquad i = 1, 2$$

where $\mathcal{F}_1 = 4.13v_r^2 + 30.34v_r + 10.35 + 2.04w_r^2 + 11.12v_r w_r + 15.2\dot{v}_r + 6\dot{w}_r$ and $\mathcal{F}_2 = (0.75 + 0.26v_r)k_1$.

We first simulate the case of tracking a line for the first 4 seconds and a circle in the remaining 5 seconds in a plane. For the line, set $v_r = 2$, $\dot{v}_r = 0$, $w_r = 0$, $\dot{w}_r = 0$ and for the circle, set $v_r = 1.4$, $\dot{v}_r = 0$, $w_r = \pi/2$, $\dot{w}_r = 0$. It is easy to check that for the line $\mathcal{F}_1 = 87.55$ and for the circle $\mathcal{F}_1 = 90.39$. The inequality $(0.75 + 2.6v_r)k_1 < 100 - \mathcal{F}_1$ under these two paths has a solution $k_1 < 8.6$. In our simulation we choose $k_1 = 5$, and get the simulation results shown in Figure 2.3. Clearly, the magnitudes of the torques are strictly less than 100 and perfect tracking is achieved. Fig.2.4 shows the estimators of disturbances.

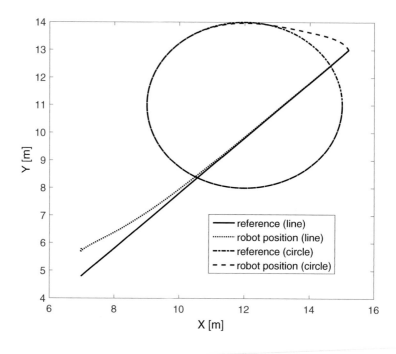

Fig. 2.2. Robot position in (x, y) plane of tracking control.

Remark 2.8. Saturation constraints of input torques restrict the controllability of systems. For a given reference, there exists a minimum saturation bound that makes perfect tracking possible. Thus for a tracking problem, certain reference trajectories cannot be tracked due to saturation constraints, regardless of the control scheme used. For example in our case, for a given reference and a saturation bound u_{Mi}, if the design parameters in Proposition 2.3 cannot guarantee $\tau_i^{\max} \leq u_{Mi}$, then the reference cannot be tracked under this saturation bound.

Secondly we consider stabilization. In this case the reference velocities are: $v_r = 0$, $w_r = 0$, so we have $|k(t)| = |\lambda_1 v_r + \lambda_2 \sin(t)| \leq |\lambda_2|$. Thus we can have a larger λ_2 to reduce the converging time compared with the case of tracking. In this case we set λ_2 to be 0.5. The initial position is set to be $\{2, 2\}$, and other parameters and initial conditions are the same as the tracking case, and we have $\tau_i^{\max} = 24.3 + 7.32k_1$. So $24.3 + 7.32k_1 < 100$ has a solution $k_1 < 10.32$. In our simulation we still choose $k_1 = 5$. The results are shown in Figure 2.7.

Thirdly we make a comparison by considering the torque design in [7], where the torque control law $\tau = B^{-1}(-k_d\tilde{v} + Y_c\hat{p} - (\frac{\partial V_1}{\partial q}\hat{S})^T)$ is developed based on combined kinematic and dynamic model without considering torque

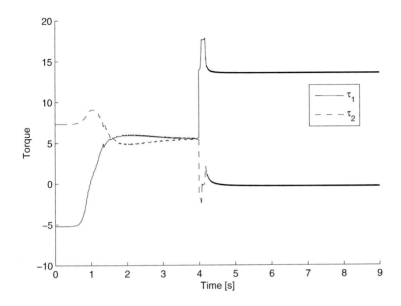

Fig. 2.3. Control torques of tracking control.

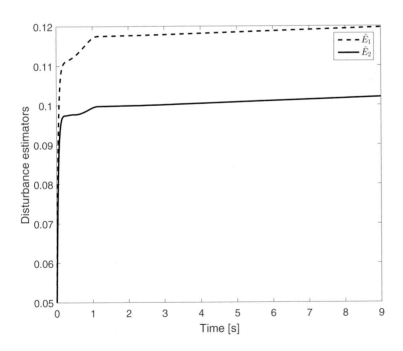

Fig. 2.4. Estimators of disturbance.

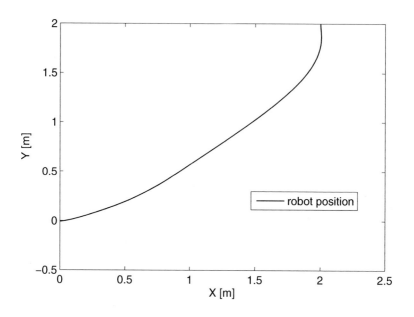

Fig. 2.5. Robot position in (x, y) plane of stabilization control.

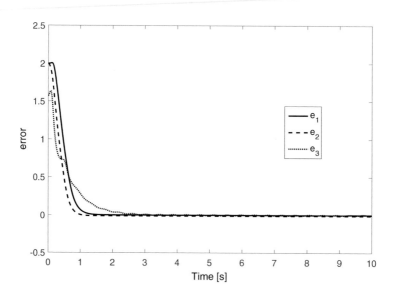

Fig. 2.6. Tracking errors of stabilization control.

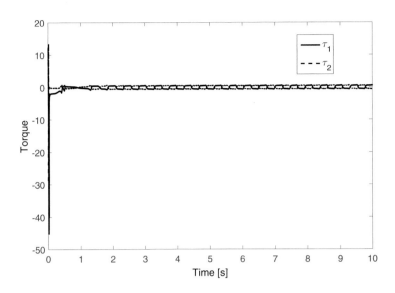

Fig. 2.7. Control torques of stabilization control.

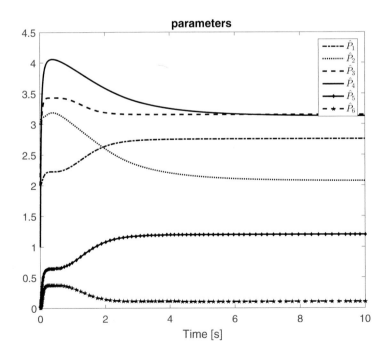

Fig. 2.8. Parameter estimators.

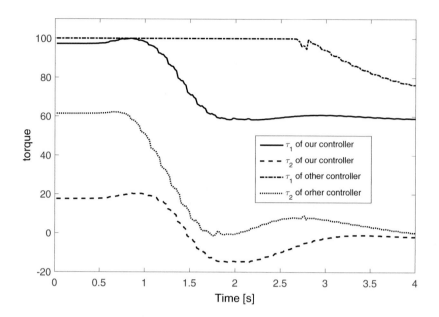

Fig. 2.9. The torque of two different controllers after suffering from input saturation.

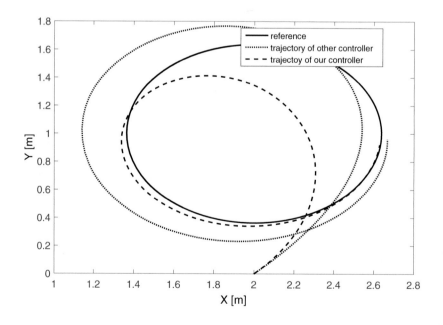

Fig. 2.10. Robot position in (x,y) plane.

saturation. The torque is applied to the system described above for the case of tracking a circle under input saturation. The results are shown in Fig.2.10, together with the tracking results obtained with our proposed tracking controller. Due to the presence of amplitude constraints, the torque applied to the system experiences saturation for the controller designed based on [7], and it is observed that perfect tracking cannot be achieved. However the amplitude constraint will not affect the tracking result of our controllers since the designed torque is always within the saturation bound.

2.5 Conclusions

A global adaptive stabilization and tracking control scheme is proposed for a nonholonomic wheeled mobile robot with input saturation and external disturbances. With the proposed scheme, we can compute the bounds of control signals in advance and thus ensure them to be within the saturation limits to avoid the occurrence of saturation by suitably choosing design parameters. System stability and perfect tracking are established. Simulation studies are also conducted to illustrate the design procedure and verify the effectiveness of the proposed scheme. However it may be possible that the computed bound of the control signal with any choice of design parameters is larger than the saturation bound. In this case, we need to explore alternative controller design methods. Other possible future work includes extending the proposed method to the controlling of other kinds of underactuated mechanical systems subject to input saturation, such as an underactuated surface vessel. Formation or flocking control of mobile robots [37, 38, 39, 40, 41] facing obstacles and communication packets loss is also an interesting area for exploration when control input is subject to saturation constraint.

2.6 Appendix

Proof of (2.28)

Let $f_2 = -\frac{v_r(\cos\bar{\phi}_e - 1)\Gamma_2}{\Gamma_1} - \frac{v_r k(t)y_e \sin\bar{\phi}_e}{\Gamma_1} = -v_r\Lambda$ where $\Lambda = \frac{(\cos\bar{\phi}_e - 1)\Gamma_2}{\Gamma_1} + \frac{k(t)y_e \sin\bar{\phi}_e}{\Gamma_1}$. We have $-\frac{y_e v_r \Lambda}{\Gamma_1} \leq \frac{\delta_1 v_r{}^2 y_e{}^2}{\Gamma_1{}^2} + \delta_1{}^{-1}\Lambda^2$ where $\delta_1 \in (0,1)$ is a positive constant. It is left to prove that $\Lambda^2 \leq \frac{\kappa\bar{\phi}_e^2}{1+\bar{\phi}_e^2}$ where κ is a positive design parameter.

Denote $f_\Lambda(x) = \Lambda_x{}^2 = [a_1(\cos(x) - 1) + a_2\sin(x)]^2$ and $f_\kappa(x) = \frac{\kappa x^2}{1+x^2}$, where $a_1 = \frac{\Gamma_2}{\Gamma_1}$, $a_2 = \frac{k(t)y_e}{\Gamma_1}$ and $0 < |a_i| < 1$ for $i = 1,2$. We first prove that there exists a positive constant κ so that $\sqrt{f_\Lambda(x)} \leq |\sqrt{f_\kappa(x)}|$ for $x \in [0, \infty)$. Define $f_4(x) = |a_1(\cos(x) - 1) + a_2\sin(x)|$ and $f_5(x) = \frac{\sqrt{\kappa}x}{\sqrt{1+x^2}}$. Note that $f_4(x)$ is periodic and attains its maximum value at $\varphi = \arcsin(\frac{|a_2|}{\sqrt{a_1{}^2 + a_2{}^2}})$.

Also $f_4(0) = 0$, $f_5(0) = 0$ and $f_4(x)$ monotonically increases in the interval for $x \in [0, \ \phi]$, it is sufficient to show $\frac{\partial f_5(x)}{\partial x} > \frac{\partial f_4(x)}{\partial x}$ for $x \in [0, \ \phi]$. Since $\frac{\partial f_4(x)}{\partial x} = \sqrt{a_1{}^2 + a_2{}^2} \sin(\varphi - x)$ for $x \in [0, \ \phi)$ and $\frac{\partial f_5(x)}{\partial x} = \frac{\sqrt{\kappa}}{(1+x^2)^{1.5}}$, it is easy to choose κ so that $\frac{\partial f_5(x)}{\partial x} > \frac{\partial f_4(x)}{\partial x}$ for $x \in [0, \ \phi]$. For example, κ could be chosen as $(1 + a_2{}^2)^3$. Letting $\delta_2 = \delta_1{}^{-1}\kappa$, then the result holds. For $x \in (-\infty, 0)$, it can be proved similarly.

3

Tracking Control of Underactuated Ships with Input Saturation

In the previous chapter, the tracking and stabilization control of nonholonomic mobile robots is investigated by normalizing certain signals and thus calculating the bounds of the input signals. However, this control scheme may not be applied to underactuated ships since the dynamic model of underactuated ships is much more complicated. In this chapter, tracking control of underactuated ships in the presence of input saturation is addressed. By dividing the tracking error dynamic system into a cascade of two subsystems, the torques in surge and yaw axis are designed separately using the backstepping technique. More specifically, we design the yaw axis torque in such a way that its corresponding subsystem is finite time stable, which makes it be de-coupled from the second subsystem after a finite time. This enables us to design the torque in the surge axis independently. It is shown that the closed-loop system is stable and the mean-square tracking errors can be made arbitrarily small by choosing design parameters.

3.1 Introduction

The control of underactuated systems with nonholonomic constraints has received vast attention over the past few decades. Typical examples of such systems include nonholonomic mobile robots, underactuated ships, underwater vehicles and VCTOL aircrafts *et al*. There has been significant interest in tracking control of underactuated ships as evidenced by various control schemes proposed for such a problem. In [17], two effective control schemes are presented to solve the global tracking control problem for underactuated ships with the Lyapunov method by exploiting the inherent cascade interconnected structure of the ship dynamics. In [42] by dividing the ship dynamics into two linear subsystems a global tracking controller is proposed. In [20] a continuous time-varying tracking controller is deigned to yield global uniformly ultimately bounded tracking by using a transformation of the ship tracking system into a skew-symmetric form and designing a time-varying dynamic oscillator. In [16] the method developed for a chain-form system was adopted for underactuated ships through a coordinate transformation to steer both the position variables and the course angle of the ship providing exponential stability of the reference trajectory. In [19] a single controller is designed to achieve stabilization and tracking of underactuated ships simultaneously. In [22] a controller that achieves practical stabilization of arbitrary reference trajectories for underactuated ships using the transverse function approach. Subsequent related works on the tracking control of underactuated ships include, but are not limited to, [21, 24, 46, 45] and many references therein.

In practice, an underactuated ship is ultimately driven by a motor which can only provide a limited amount of torque. Thus the magnitude of the control signal is always constrained. Under such a constraint, the controllability of the system is seriously affected. Thus it often severely limits system performance, giving rise to undesirable inaccuracy or leading to instability. This brings great challenges to the design and analysis of controllers. Therefore, the development of a control scheme for an underactuated ship with input saturation has been an extremely difficult task, yet with practical interest and theoretical significance. In [47] a tracking control method is presented to address such an issue by employing the linearization and dynamic surface control, only ensuring bounded tracking errors. Other than this one, there is no other literature dealing with this problem as far as we know, due to its difficulty.

In this chapter, we solve this problem by proposing a state-feedback control scheme. Firstly the tracking error dynamic system is divided into a cascade of two subsystems so that the torques in surge and yaw axis can be designed separately based on the two subsystems. A new finite-time control scheme is proposed to design the torque in yaw which enables the two subsystems to be de-coupled within finite time. After decoupling, a new backstepping scheme similar to that in [48] is proposed for the surge subsystem. The designed controllers for the two subsystems are analyzed, respectively. Then the overall system is studied. It is shown that the closed-loop system is stable and the mean-square tracking errors can be made arbitrarily small by adjusting design parameters. To illustrate the effectiveness of the controller, simulation studies are conducted.

3.2 Problem Formation

3.2.1 Underactuated Ship Model

Similar to [17], we consider the ship as shown in Figure 3.1. The only two propellers are the torques in surge and yaw axis. The motion of the ship dynamics is described by the following differential equations

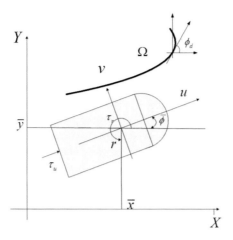

Fig. 3.1. The coordinates of an underactuated ship.

$$\dot{x} = u \cos \phi - v \sin \phi$$
$$\dot{y} = u \sin \phi + v \cos \phi$$
$$\dot{\phi} = r \tag{3.1}$$
$$\dot{u} = \frac{m_2}{m_1} vr - \frac{d_1}{m_1} u + \frac{1}{m_1} Sat_u(\tau_u)$$
$$\dot{v} = -\frac{m_1}{m_2} ur - \frac{d_2}{m_2} v$$
$$\dot{r} = \frac{m_1 - m_2}{m_3} uv - \frac{d_3}{m_3} r + \frac{1}{m_3} Sat_r(\tau_r) \tag{3.2}$$

where (x, y) denotes the position of the ship surface in X and Y directions, respectively, ϕ denotes the heading angle of the ship, u, v and r are the surge, sway and yaw velocities, respectively. τ_u and τ_r represent the torques in surge and yaw axis, respectively, which are considered as actual control inputs. Since there is no torque introduced in the sway direction, the system modeled by (3.1)-(3.2) is underactuated. The positive constants m_j denote the ship inertia including added mass, and d_j, $1 \leq j \leq 3$ represent the hydrodynamic damping coefficients. The saturation function is defined as follows

$$Sat_\star(x) = \begin{cases} x & \text{if } |x| < \tau_{\star \max} \\ \tau_{\star \max} & \text{if } |x| \geq \tau_{\star \max} \end{cases} \tag{3.3}$$

where $\star$ denotes u or r respectively and $\tau_{u \max}$ and $\tau_{r \max}$ are their respective saturation limits.

Our control objective is formally stated as follows.

- *Control Objective*: Design the control inputs τ_u and τ_r to force the ship, as modeled in (3.1)-(3.2) subject to input saturation, to track a prescribed path, which is denoted by $\Omega = (x_d, y_d, \phi_d, u_d, r_d)$ generated by the dynamic equations of a virtual underactuated ship:

$$\dot{x}_d = u_d \cos(\phi_d) - v_d \sin(\phi_d)$$
$$\dot{y}_d = u_d \sin(\phi_d) + v_d \cos(\phi_d)$$
$$\dot{\phi}_d = r_d$$
$$\dot{v}_d = -\frac{m_{11}}{m_{22}} u_d r_d - \frac{d_2}{m_2} v_d. \tag{3.4}$$

3.2.2 Variable Transformation

To facilitate the control design, firstly we define the path-following errors $\{x_e, y_e, \phi_e\}$ as follows

$$\begin{bmatrix} x_e \\ y_e \\ \phi_e \end{bmatrix} = \begin{bmatrix} \cos\phi & \sin\phi & 0 \\ -\sin\phi & \cos\phi & 0 \\ 0 & 0 & 1 \end{bmatrix} \begin{bmatrix} x - x_d \\ y - y_d \\ \phi - \phi_d \end{bmatrix}. \tag{3.5}$$

Differentiating both sides of (3.5) along (3.1) results in the kinematic error dynamics:

$$\dot{x}_e = u_e - u_d\left(\cos(\phi_e) - 1\right) - v_d \sin(\phi_e) + r_e y_e + r_d y_e$$
$$\dot{y}_e = v_e - v_d\left(\cos(\phi_e) - 1\right) + u_d \sin(\phi_e) - r_e x_e - r_d x_e$$
$$\dot{\phi}_e = r_e$$
$$\dot{u}_e = \frac{m_2}{m_1} vr - \frac{d_1}{m_1} u + \frac{1}{m_1} Sat_u(\tau_u) - \dot{u}_d$$
$$\dot{v}_e = -\frac{m_1}{m_2}(ur - u_d r_d) - \frac{d_2}{m_2} v_e$$
$$\dot{r}_e = \frac{m_1 - m_2}{m_3} uv - \frac{d_3}{m_3} r + \frac{1}{m_3} Sat_r(\tau_r) - \dot{r}_d \tag{3.6}$$

where $u_e = u - u_d$, $v_e = v - v_d$ and $r_e = r - r_d$.

To achieve the control objective, the following assumption is imposed.

Assumption 3.1 *There is a constant $\delta_r > 0$ such that, for any time period (t_0, t),*

$$\int_{t_0}^{t} r_d^2(\tau) d\tau \geq \delta_r(t - t_0).$$

Assumption 3.2 *There exists a positive constant Δ such that*

$$\frac{|m_1 - m_2|}{2d_1\alpha} \tau_{u\,\mathrm{max}}^2 + \Delta \leq \tau_{r\,\mathrm{max}} \tag{3.7}$$

where $\alpha = \min\{\frac{d_1}{m_1}, \frac{2d_2}{m_2}\}$. Furthermore, r_d is chosen such that $|m_d \dot{r}_d + d_3 r_d| < \Delta$.

Remark 3.1. Assumption 3.1 is a PE condition on the reference angular velocity r_d which is commonly required, see for example [17, 42] etc. Assumption 3.2 is the restrictions for the saturation limits and the reference that is needed for controller design.

3.3 Controller Design

The control design procedure can be summarized as follows. Firstly (3.6) will be divided into a cascade of two subsystems, i.e., (ϕ_e, r_e) subsystem and (x_e, y_e, v_e, u_e) subsystem. Control τ_r is designed such that ϕ_e and r_e converge to zero within finite time T, while it guarantees that $|\tau_r| < \tau_{r\max}$. For $t > T$ the two subsystems are de-coupled since the interaction effects are only from (ϕ_e, r_e) subsystem to (x_e, y_e, v_e, u_e) subsystem, i.e. one-way interaction. Control τ_u is designed based on (x_e, y_e, v_e, u_e) subsystem with a new backstepping technique in such a way that τ_u is the output of a filter. It is then shown that the overall closed-loop system is stable and the mean-square tracking errors could be arbitrarily small by adjusting design parameters.

Now we develop the following intermediate result to reveal the bounds of u, v and r.

Lemma 3.2. *The variables u, v and r are bounded and their respective bounds satisfy*

$$\limsup_{t \to \infty} |u| \leq \sqrt{\frac{m_2}{2m_1 d_1 \alpha}} \tau_{u\max}, \quad |v| \leq \sqrt{\frac{m_1}{2m_2 d_1 \alpha}} \tau_{u\max},$$
$$|r| \leq \frac{\tau_{u\max}}{\sqrt{m_3 \beta d_1}} + \frac{\tau_{r\max}}{\sqrt{m_3 \beta d_3}} \tag{3.8}$$

where $\beta = \min\{\frac{d_1}{m_1}, \frac{2d_2}{m_2}, \frac{d_3}{m_3}\}$.

Proof: Consider a Lyapunov function $V_0 = \frac{m_1}{2m_2}u^2 + \frac{m_2}{2m_1}v^2$, whose derivative is

$$\dot{V}_0 = -\frac{d_1}{m_2}u^2 - \frac{d_2}{m_1}v^2 + \frac{1}{m_2}u Sat_u(\tau_u)$$
$$\leq -\frac{d_1}{m_2}u^2 - \frac{d_2}{m_1}v^2 + \frac{d_1}{2m_2}u^2 + \frac{1}{2d_1}\tau_{u\max}^2$$
$$\leq -\alpha V_0 + \frac{1}{2d_1}\tau_{u\max}^2.$$

Without loss of generality, let $u(0) = 0$ and $v(0) = 0$. Thus $V_0 \leq \frac{\tau_{u\max}^2}{2d_1 \alpha}$. So $|u| \leq \sqrt{\frac{m_2}{2m_1 d_1 \alpha}}\tau_{u\max}$ and $|v| \leq \sqrt{\frac{m_1}{2m_2 d_1 \alpha}}\tau_{u\max}$. For r, consider $V = \frac{m_1}{2}u^2 + \frac{m_2}{2}v^2 + \frac{m_3}{2}r^2$, whose derivative is

$$\dot{V} = -d_1 u^2 - d_2 v^2 - d_3 r^2 + uSat_u(\tau_u) + rSat_r(\tau_r)$$

$$\leq -\frac{d_1}{2}u^2 - d_2 v^2 - \frac{d_3}{2}r^2 + \frac{\tau_{u\,\max}^2}{2d_1} + \frac{\tau_{r\,\max}^2}{2d_3}$$

$$\leq -\beta V + \frac{\tau_{u\,\max}^2}{2d_1} + \frac{\tau_{r\,\max}^2}{2d_3}.$$

So $V \leq \frac{\tau_{u\,\max}^2}{2\beta d_1} + \frac{\tau_{r\,\max}^2}{2\beta d_3}$ and $|r| \leq \sqrt{\frac{\tau_{u\,\max}^2}{m_3\beta d_1} + \frac{\tau_{r\,\max}^2}{m_3\beta d_3}} \leq \frac{\tau_{u\,\max}}{\sqrt{m_3\beta d_1}} + \frac{\tau_{r\,\max}}{\sqrt{m_3\beta d_3}}$. This completes the proof. □

Now we design τ_r with the aim that $\sup|\tau_r| \leq \tau_{r\,\max}$ and ϕ_e and r_e defined in (3.6) converge to zero within finite time. Consider the following system

$$\dot{\phi}_e = r_e$$

$$\dot{r}_e = \frac{m_1 - m_2}{m_3}uv - \frac{d_3}{m_3}r + \frac{1}{m_3}Sat_r(\tau_r) - \dot{r}_d. \tag{3.9}$$

First of all, define $sig(x)^k = sign(x)|x|^k$, then τ_r is designed as

$$\tau_r = (m_2 - m_1)uv + m_3\dot{r}_d + d_3 r_d$$
$$- m_3 Sat_{\varepsilon_1}\left(sig(\phi_e)^{k_1}\right) - m_3 Sat_{\varepsilon_2}\left(sig(r_e)^{k_2}\right) \tag{3.10}$$

where ε_1 and ε_2 are positive design parameters, k_1 and k_2 are two positive constants satisfying $0 < k_1 < 1$, $k_2 = \frac{2k_1}{1 + k_1}$.

Substituting (3.10) into (3.9),

$$\dot{\phi}_e = r_e$$

$$\dot{r}_e = -Sat_{\varepsilon_1}\left(sig(\phi_e)^{k_1}\right) - Sat_{\varepsilon_2}\left(sig(r_e)^{k_2}\right) - \frac{d_3}{m_3}r_e. \tag{3.11}$$

Then the following lemma is established.

Lemma 3.3. *With controller (3.10), ϕ_e and r_e in (3.11) will converge to zero within finite time. Furthermore, it is ensured that $|\tau_r| \leq \tau_{r\,\max}$ $\forall t$ by choosing appropriate design parameters ε_1 and ε_2.*

Proof: We first define V_x as

$$V_x = \begin{cases} \frac{1}{1+k_1}|x|^{1+k_1} & |x| \leq \varepsilon_1^{1/k_1} \\ \frac{1}{1+k_1}|\varepsilon_1|^{\frac{1+k_1}{k_1}} + sgn(x)\varepsilon_1\left(x - sgn(x)\varepsilon_1^{1/k_1}\right) & |x| > \varepsilon_1^{1/k_1} \end{cases} \tag{3.12}$$

which is plotted in Fig. 3.2. It is obvious that V_x is positive semi-definite, unbounded and is C^1. Also $\dot{V}_{\phi_e} = sat_{\varepsilon_1}(sig(\phi_e)^{k_1})\dot{\phi}_e$. Then consider the following Lyapunov function

$$V_1 = V_{\phi_e} + \frac{1}{2}r_e^2, \tag{3.13}$$

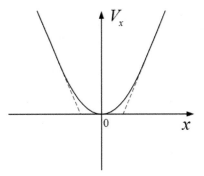

Fig. 3.2. Shape of V_x.

whose derivative along (3.11) is

$$\dot{V}_1 = -\frac{d_3}{m_3} r_e^2 - r_e sat_{\varepsilon_2}(sig(r_e)^{k_2}). \tag{3.14}$$

From (3.14), r_e is globally asymptotically convergent. Similarly it can also be proved that ϕ_e is globally asymptotically convergent. Note that the saturation in (3.11) will not take effect when

$$|\phi_e| \le \varepsilon_1^{1/k_1}, \quad |r_e| \le \varepsilon_2^{1/k_2}. \tag{3.15}$$

Thus we define a set

$$\Pi = \{(\phi_e, r_e) : \frac{1}{1+k_1}|\phi_e|^{1+k_1} + \frac{1}{2}r_e^2 \le \theta\} \tag{3.16}$$

where $\theta = \min\{\frac{\varepsilon_1^{(1+k_1)/k_1}}{1+k_1}, \frac{1}{2}\varepsilon_2^{2/k_2}\}$. Now consider $V_2 = \frac{1}{1+k_1}|\phi_e|^{1+k_1} + \frac{1}{2}r_e^2$. When $\phi_e \in \Pi$ and $r_e \in \Pi$, the saturation does not take effect, and it can be easily checked that $\dot{V}_2 = -\frac{d_3}{m_3}r_e^2 - |r_e|^{1+k_2} \le 0$, thus Π is an invariant set.

When $(\phi_e, r_e) \notin \Pi$, (ϕ_e, r_e) approaches Π within finite time from (3.14), and after (ϕ_e, r_e) reaches Π, we have

$$\dot{\phi}_e = r_e$$
$$\dot{r}_e = -\frac{d_3}{m_3}r_e - \left(sig(\phi_e)^{k_1} + sig(r_e)^{k_2}\right). \tag{3.17}$$

Next we show that (3.17) is finite-time stable. To establish the stability of (3.17), the following fact is presented.

Fact 3.1 *The following system is globally finite-time stable.*

$$\dot{x} = y \tag{3.18a}$$
$$\dot{y} = -sig(x)^{k_1} - sig(y)^{k_2} \tag{3.18b}$$

where k_1 and k_2 are two positive constants satisfying $0 < k_1 < 1$, $k_2 = \dfrac{2k_1}{1+k_1}$.

Based on Fact 3.1,

$$\dot{\phi}_e = r_e$$
$$\dot{r}_e = -sig(\phi_e)^{k_1} - sig(r_e)^{k_2} \tag{3.19}$$

is globally finite-time stable, and is homogeneous of degree $\kappa = k_1 - 1 < 0$ with respect to $(r_1, r_2) = (\frac{2}{1+k_1}, 1)$. Furthermore

$$\lim_{\varepsilon \to 0} \frac{d_3 \varepsilon^{r_2} r_e}{m_3 \varepsilon^{\kappa + r_2}} = \frac{d_3}{m_3} r_e \lim_{\varepsilon \to 0} \varepsilon^{-\kappa} = 0 \tag{3.20}$$

which means $-\frac{d_3}{m_3} r_e$ is 'higher degree' with respect to $\varepsilon^{\kappa + r_2}$ in the sense of (3.20). Based on Lemma 3.11 in the Appendix, we know that (3.17) is finite-time stable if $(\phi_e, r_e) \in \Pi$.

So in summary, if $(\phi_e(0), r_e(0)) \notin \Pi$, (ϕ_e, r_e) will reach Π within finite time, and when $(\phi_e, r_e) \in \Pi$, (ϕ_e, r_e) will converge to the origin within finite time. Thus (ϕ_e, r_e) will approach to zero within a total of finite time T and therefore(3.11) is globally finite time stable.

Also from (3.10) and Lemma 3.2,

$$|\tau_r| \le |m_2 - m_1| \frac{1}{2d_1 \alpha} \tau_{u\,\max}^2 + m_3(\varepsilon_1 + \varepsilon_2) + \left| m_3 \dot{r}_d + d_3 r_d \right|.$$

Based on Assumption 3.1, choose ε_1 and ε_2 to satisfy

$$\left| m_3 \dot{r}_d + d_3 r_d \right| + m_3(\varepsilon_1 + \varepsilon_2) \le \Delta. \tag{3.21}$$

Then we have

$$|\tau_r| \le \tau_{r\,\max} \tag{3.22}$$

which means the saturation does not take effect for the yaw subsystem. This completes the proof. $\qquad\square$

Since ϕ_e and r_e will converge to zero within a finite time T, and due to the saturation, u, v and r are bounded, thus for $t \le T$, x_e, y_e, ϕ_e, u_e and r_e are bounded. As for $t > T$, $\phi_e = 0$ and $r_e = 0$, then the (x_e, y_e, v_e, u_e) subsystem becomes

$$\dot{y}_e = v_e - r_d x_e$$
$$\dot{x}_e = u_e + r_d y_e$$
$$\dot{v}_e = -\frac{m_1}{m_2} u_e r_d - \frac{d_2}{m_2} v_e$$
$$\dot{u}_e = \frac{m_2}{m_1} v r_d - \frac{d_1}{m_1} u + \frac{1}{m_1} Sat_u(\tau_u) - \dot{u}_d. \tag{3.23}$$

The model in (3.23) enables us to design τ_u for $t > T$ without considering the coupling effects from (ϕ_e, r_e) subsystem. Due to the one-way interaction, the

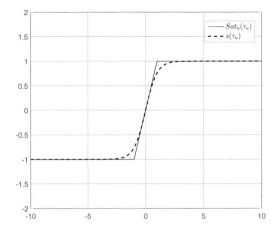

Fig. 3.3. Approximation of the saturation function $Sat_1(\tau_u)$.

designed control τ_u can still be applied from the initial time $t = 0$. As $Sat_u(\cdot)$ is non-smooth, thus backstepping technique, see [8], cannot be applied here. To overcome this difficulty, the saturation will be approximated by a smooth function defined as in [48]

$$s(\tau_u) = \tau_{u\,\max} \tanh(\tau_u/\tau_{u\,\max}) = \tau_{u\,\max} \frac{e^{\tau_u/\tau_{u\,\max}} - e^{-\tau_u/\tau_{u\,\max}}}{e^{\tau_u/\tau_{u\,\max}} + e^{-\tau_u/\tau_{u\,\max}}}. \qquad (3.24)$$

Then $Sat_u(\cdot)$ can be expressed as

$$Sat_u(\tau_u) = s(\tau_u) + d(t) \qquad (3.25)$$

where $d(t) = Sat_u(\tau_u) - s(\tau_u)$ is a bounded function which can be deemed as a 'disturbance' bounded as

$$|d(t)| = |Sat_u(\tau_u) - s(\tau_u)| \le \tau_{u\,\max}(1 - \tanh(1)) := D. \qquad (3.26)$$

Figure 3.3 shows an example of approximating the saturation function. To achieve the control objective, we augment the closed-loop system (3.23) by including the saturation approximation function and the resulting approximation error as follows

$$\dot{y}_e = v_e - r_d x_e$$
$$\dot{x}_e = u_e + r_d y_e$$
$$\dot{v}_e = -\frac{m_1}{m_2} u_e r_d - \frac{d_2}{m_2} v_e$$
$$\dot{u}_e = \frac{m_2}{m_1} v r_d - \frac{d_1}{m_1} u + \frac{1}{m_1} s(\tau_u) - \dot{u}_d + \frac{1}{m_1} d(t)$$
$$\dot{\tau}_u = -c\tau_u + \alpha \qquad (3.27)$$

Virtual control errors:

$$\tilde{x}_e = x_e - \alpha_{y_e}$$
$$\tilde{u}_e = u_e - \alpha_{x_e}$$
$$\tilde{\tau}_u = \alpha_{\tau_u} - s(\tau_u) \qquad (3.28)$$

Virtual Controllers and Intermediate Controller:

$$\alpha_{y_e} = c_1 r_d y_e$$
$$\alpha_{x_e} = -c_2 \tilde{x}_e$$
$$\alpha_{\tau_u} = -c_4 \tilde{u}_e - m_2 v r_d + d_1 u + m_1(\dot{u}_d + \dot{\alpha}_{x_e}$$
$$-\tilde{x}_e) + \frac{m_1^2 c_3}{m_2} v_e r_d - \tanh(\frac{1}{\varepsilon}\tilde{u}_e)\frac{D}{m_1}$$
$$\alpha = N(\chi)\bar{\alpha}$$
$$\bar{\alpha} = -c_5\tilde{\tau}_u + c\tau_u\frac{\partial s(\tau_u)}{\partial \tau_u} + \dot{\alpha}_{\tau_u} - \frac{1}{m_1}\tilde{u}_e$$
$$N(\chi) = \chi^2\cos(\frac{\pi\chi}{2}), \quad \dot{\chi} = \bar{\alpha}\tilde{\tau}_u \qquad (3.29)$$

Condition for design parameters:

$$c_2 - c_1|r_d|^2 > 0,$$
$$-\tilde{x}_e\left(c_1\dot{r}_d y_e + c_1 r_d(v_e - c_1 r_d^2 y_e)\right) + y_e v_e + \frac{c_2 c_3 m_1}{m_2} r_d v_e \tilde{x}_e$$
$$\leq (1-\epsilon)(c_2 - c_1|r_d|^2)\tilde{x}_e^2 - \frac{m_2}{(1-\epsilon)d_2 c_3}y_e^2 - \frac{(1-\epsilon)c_3 d_2}{m_2}v_e \qquad (3.30)$$

Table 3.1. Summary of virtual control errors, virtual controls, intermediate control and design parameter condition.

where c is a positive constant and α is an auxiliary intermediate control to be designed in the backstepping approach. Clearly all functions in (3.27) are smooth and the use of backstepping is feasible. Let α_{y_e} be the virtual controls of x_e to stabilize y_e, α_{x_e} be the virtual control of u_e and α_{τ_u} be the virtual control of $s(\tau_u)$. Then following the standard backstepping design steps as in [8], these virtual controls and the intermediate control α in (3.27) can be obtained as summarized in Table 3.1 where $0 < \epsilon < 1$ is a positive constant and $N(\chi)$ is a Nassbaum function. From (3.30) the choices of design parameters c_1 and c_2 are dependent on $|r_d|$ and $|\dot{r}_d|$, and are possible. Similar analysis of choosing c_1 and c_2 can also be found in [17].

The final control τ_u is generated from the output of the filter

$$\dot{\tau}_u = -c\tau_u + \alpha \tag{3.31}$$

in (3.27).

Remark 3.4. Although τ_u is designed based on the de-coupled surge subsystem in (3.27) for $t > T$, it is implemented for all $t \geq 0$. It has no effects on (ϕ_e, r_e) subsystem and thus does not affect the results in Lemma 3.3.

Consider the following Lyapunov function

$$V_3 = \frac{1}{2}y_e^2 + \frac{1}{2}\tilde{x}_e^2 + \frac{c_3}{2}v_e^2 + \frac{1}{2}\tilde{u}_e^2 + \frac{1}{2}\tilde{\tau}_u^2. \tag{3.32}$$

Then we have

$$\begin{aligned}
\dot{V}_3 &\leq (\xi N(\chi) - 1)\bar{\alpha}\tilde{\tau}_u - \tilde{c}V_3 + D\tilde{u}_e f(\tilde{u}_e) \\
&= (\xi N(\chi) - 1)\dot{\chi} - \tilde{c}V_3 + O(\varepsilon)
\end{aligned} \tag{3.33}$$

where $O(\varepsilon) = \frac{D}{m_1}x\left(sign(x) - \tanh(\frac{1}{\varepsilon}x)\right)$ for $x \in \Re$ and $\tilde{c} = \min\{c_1 r_d^2 - \frac{m_2}{(1-\epsilon)d_2 c_3}, \epsilon(c_2 - c_1 r_d^2), \frac{\epsilon c_3 d_2}{m_2}, 2c_4, 2c_5\} > 0, \xi = \frac{\partial s(\tau_u)}{\partial \tau_u} = \frac{4}{(e^{\tau_u/\tau_u \max} + e^{-\tau_u/\tau_u \max})^2} \in (0,1)$. It should be noted that $\sup_{\forall x} O(\varepsilon)$ is arbitrarily small if ε is small enough.

With the proposed control, the following lemma is established.

Lemma 3.5. *Consider system (3.27) under the control of (3.28)-(3.30). If Assumption 3.1 and Assumption 3.2 are satisfied and $V_3(T)$ is bounded, then all signals in (3.27) are bounded for $t > T$.*

Proof:
From (3.33) we have

$$V_3(t) \leq V_3(T)e^{-\tilde{c}(t-T)} + \frac{O(\varepsilon)}{\tilde{c}}(1 - e^{-\tilde{c}(t-T)}) + e^{-\tilde{c}t}\int_T^t (\xi N(\chi)\dot{\chi} - \dot{\chi})e^{\tilde{c}\sigma}d\sigma \tag{3.34}$$

for $t > T$. The boundedness of V_3 can be established based on the Nussbaum properties via a contradiction argument. First define V_N on the time-period (t_i, t_j) as

$$V_N(t_i, t_j) = \int_{t_i}^{t_j} (\xi N(\chi) - 1)\dot{\chi}e^{-\tilde{c}(t_j - \sigma)}d\sigma. \tag{3.35}$$

For notation convenience, let $V_N(t_i, t_j) = V_N(\chi_i, \chi_j), T < t_i < t_j$. By noting the fact that $0 < \xi < 1$ and $0 < e^{-\tilde{c}(t_j - \sigma)} < 1$, we have

$$|V_N(\chi_i, \chi_j)| \leq \int_{\chi_i}^{\chi_j} (|N(\chi)| + 1) d\chi$$

$$\leq (\chi_j - \chi_i) \left(sup_{\chi \in [\chi_i, \chi_j]} |N(\chi)| + 1 \right). \tag{3.36}$$

For the Nussbaum function $N(\chi) = \chi^2 \cos(\frac{\pi \chi}{2})$, we know it is positive for $\chi \in (4m - 1, 4m + 1)$ and negative for $\chi \in (4m + 1, 4m + 3)$ with an integer m. Thus we consider two time periods: $[\chi_0, \chi_1] = [\chi_0, 4m + 1]$ and $[\chi_1, \chi_2] = [4m+1, 4m+3]$ for $\chi_0 > 0$ and m being a positive integer such that for $t \to t_f$, where t_f is an arbitrary positive constant, we have $\chi = \chi_2$. From (3.36) we know for $[\chi_0, \chi_1] = [\chi_0, 4m + 1]$, and we have

$$V_N(\chi_0, \chi_2) = V_N(\chi_0, \chi_1) + V_N(\chi_1, \chi_2)$$

$$\leq (4m + 1)^2 \Big(- l_2((8m + 2)(1 - \Psi) + (1 - \Psi)^2)$$

$$+ 4m + 1 - \chi_0 + \frac{l_1 + l + 3}{(4m + 1)^2} \Big) \tag{3.37}$$

where $l_2 = 2\Psi e^{-\tilde{c}*(t_2-t_1)} \cos(\pi \Psi/2) > 0$ and $l_3 = 2\Psi e^{-\tilde{c}*(t_2-t_1)} > 0$.

We now prove χ is bounded over (T, t_f) by seeking a contradiction. Suppose χ is unbounded over (T, t_f). Two cases should be considered: (i) χ has no upper-bound and (ii) χ has no lower-bound. If $\chi(t)$ has no upper bound on (T, t_f), in this case, there must exist a time interval $[t_s, t_f]$ such that $\chi(t_s) > 0$, when $t \to t_f$, $\chi(t) = \infty$. However, from (3.37) we know $V_N(\chi_0, \chi_2) \to -\infty$ as $m \to \infty$. On the other hand, $V_3(t) > 0$ for all t. Thus we can always find a subsequence of the interval that leads to a contradiction. So $\chi(t)$ has a upper bound. On the other side, if $\chi(t)$ has no lower bound on (T, t_f). Define $\chi = -\lambda$, then λ has no upper bound accordingly. Since $N(\chi)$ is an even function, thus (3.33) becomes

$$V_3(t) \leq V_3(T) e^{-\tilde{c}(t-T)} + \frac{O(\varepsilon)}{\tilde{c}} (1 - e^{-\tilde{c}(t-T)})$$

$$- \int_T^t (\xi N(\lambda) - 1) \dot{\lambda} e^{-\tilde{c}(t-\sigma)} d\sigma$$

$$= V_3(T) e^{-\tilde{c}(t-T)} + \frac{O(\varepsilon)}{\tilde{c}} (1 - e^{-\tilde{c}(t-T)}) - V_N(\lambda(T), \lambda(t)). \tag{3.38}$$

Therefore there exists a time interval $[t_s, t_f]$ such that $\lambda(t)$ is monotonically increasing and $\lim_{t \to t_f} \lambda(t) = \infty$ with $\lambda(t_s) > 0$. Following the same procedure, we can construct a subsequence of the interval that leads to a contradiction. Therefore χ has a lower bound on (T, t_f). Then from (3.36) V_N is bounded. The above argument holds for all $t_f > T$. So χ is bounded. Thus V_3 is also bounded, which means x_e, y_e, v_e, $\tilde{u}_e$ and $\tilde{\tau}_u$ are bounded. This completes the proof.   □

Remark 3.6. By examining the methodologies of stability analysis using backstepping approaches in existing literatures, it is noted that control signals are shown bounded after establishing the boundedness of virtual control signals. Based on such an idea, we can prove $s(\tau_u)$ bounded after showing the boundedness of $\tilde{\tau}_u$ and α_{τ_u} in (3.28). However, this is not helpful at in our case a $s(\tau_u)$ itself is always bounded. In other words, the boundedness of τ_u cannot be obtained from the boundedness of function $s(\tau_u)$. So such existing methodologies are not applicable here. This is the main challenge encountered to address the issue of input saturation. To overcome the challenge, the Nassbaum technique is applied here to establish the boundedness of the intermediate control α and then the boundedness of τ_u. However, due to the difficulty of applying Nassbaum to MIMO nonlinear systems, the ship dynamic system is de-coupled to two single-input subsystems with one finite-time controller designed for τ_r.

With Lemma 3.5, the following theorem is established.

Theorem 3.7. *Consider the closed-loop system consisting of the underactuated ship dynamic systems (3.1)-(3.2) and the proposed controllers (3.10), (3.28)-(3.30) and (3.31) designed under Assumption 3.1 and Assumption 3.2. The orientational error ϕ_e converges to origin within finite time. The mean-square of position errors $x_e(t)$ and $y_e(t)$ are of the order of ε where ε is a design parameter..*

Proof: We consider $t \in (0, T)$. Since τ_u and τ_r are saturated, it is easy to be checked that u, v, r, x, y, ϕ, $\tilde{u}_e$, v_e, r_e, x_e, y_e, ϕ_e and their derivatives are bounded for $t \in (0, T)$. From (3.29) α_{τ_u} and $\dot{\alpha}_{\tau_u}$ are bounded. Thus from (3.28) we know $\tilde{\tau}_u$ is bounded. Then based on (3.27) and (3.29), we know τ_u and χ are bounded for $t \in (0, T)$. Thus from Lemma 3.5 we know the closed-loop system is stable and the orientational error ϕ_e converges to origin within finite time T.

From (3.33),

$$\int_0^{T_1} V_3(t)dt \leq \int_0^T V_3(t)dt + \frac{V_3(T_1) - V_3(T)}{\tilde{c}}$$
$$+ \frac{1}{\tilde{c}} \int_{\chi(T)}^{\chi(T_1)} (\xi N(\chi) - 1)d\chi + \frac{1}{\tilde{c}} \int_T^{T_1} O(\varepsilon)dt. \qquad (3.39)$$

Since $\int_0^T V_3(t)dt$ and $\int_{\chi(T)}^{\chi(T_1)}(\xi N(\chi) - 1)d\chi$ are bounded, we have

$$\lim_{T_1 \to \infty} \frac{1}{T_1} \int_0^{T_1} V_3(t)dt \leq \frac{O(\varepsilon)}{\tilde{c}} \qquad (3.40)$$

which shows that the mean-square of errors x_e, y_e, v_e, $\tilde{u}_e$ and $\tilde{\tau}_u$ are of the order of ε. So the mean-square tracking errors could be arbitrarily small in the sense of (3.40) by adjusting ε. This completes the proof. □

Remark 3.8. Although the control scheme to deal with input saturation is developed for underactuated ship, but the methodology can be applied to other mechanical systems, e.g. nonholonomic mobile robots.

3.4 Simulations

In this section, we carry out simulations to demonstrate the effectiveness of our controllers and to validate our constructive methodology for underactuated ships. The parameters of the underactuated ship from [17] are used, namely $m_1 = 0.1$, $m_2 = 0.1$, $m_3 = 0.1$, $d_1 = 0.1$, $d_2 = 0.2$ and $d_3 = 0.1$. The saturation limits are given as $\tau_{u\max} = 0.05$ and $\tau_{r\max} = 0.12$. The reference velocities are chosen as $u_d = 0.1$ and $r_d = 0.1$, respectively. The initial position of the reference ship is given to be $(x_d(0), y_d(0), \phi_d(0), v_d(0)) = (0, 0, 0, 0)$. For the real ship, we consider the following initial position: $(x(0), y(0),$ $\phi(0), u(0), v(0), r(0)) = (0.1, 0.2, 0.1, 0, 0, 0)$. The control parameters are selected as: $c_1 = 4$, $c_2 = 2$, $c_3 = 10$, $c_4 = 4$, $c_5 = 3$, $c = 2$, $\varepsilon_1 = 0.6$, $\varepsilon_2 = 0.6$, $\varepsilon = 0.05$. Simulation time is 40s.

Figure 3.4 shows the reference and the trajectory of the real ship. Figure 3.5 shows the position errors and orientational error. It can be seen that the reference can be tracked with arbitrarily small error as governed by ε. Figure 3.6 shows the torque τ_u and τ_r. It is obvious that the maximum value of $|\tau_u|$ and $|\tau_r|$ are less than $\tau_{u\max}$ and $\tau_{r\max}$ respectively.

As a comparison, the control scheme from [17] without considering saturation is also applied to the above system subject to the same saturation limits and with the same reference and initial conditions. The results are shown in Fig 3.7. It can be observed that the controllers designed with our proposed scheme perform better in tracking the reference.

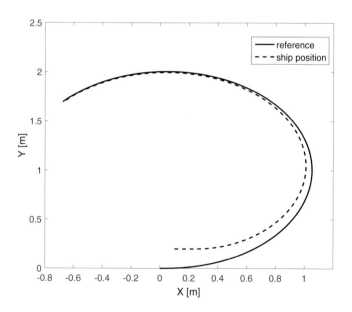

Fig. 3.4. The positions of the reference and real ship.

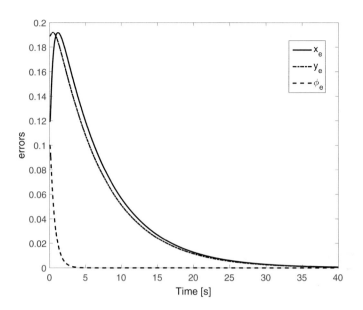

Fig. 3.5. The position errors x_e, y_e and the orientational error ϕ_e of the real ship.

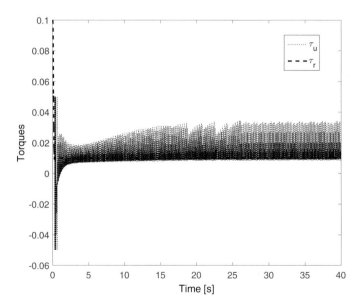

Fig. 3.6. The torque τ_u and τ_r.

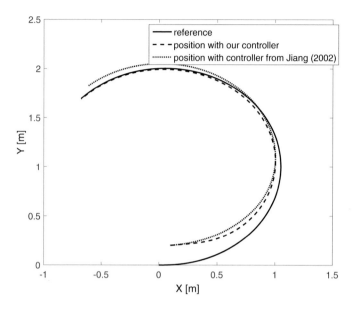

Fig. 3.7. The position of the reference, real ship with our control scheme and real ship with control technique from [17] subject to saturation.

3.5 Conclusion

In this chapter, we consider global tracking control problem of underactuated ship subject to input saturation. First the tracking error dynamic system is divided into a cascade of two subsystems, i.e., (ϕ_e, r_e) subsystem and (x_e, y_e, v_e, u_e) subsystem. With our proposed finite-time yaw controller, the two subsystems are de-coupled within finite time T. This enables the torque in surge to be designed independently by using a new backstepping technique. It is shown that the system is stable and the mean-square tracking errors can be made arbitrarily small by adjusting the design parameters.

3.6 Appendix

We will introduce some concepts of homogeneous functions following [43, 44].

Definition 3.9. *Suppose* $(r_1, ..., r_n) \in R^n$ *where* $r_i > 0$, $i = 1, ..., n$, *and* $V(x_1, ..., x_n) : R^n \to R$ *being a continuous function. Then* $V(x_1, ..., x_n)$ *is homogeneous of degree* $\sigma > 0$ *with respect to* $(r_1, ..., r_n)$, *if for any given* ε,

$$V(\varepsilon^{r_1} x_1, ..., \varepsilon^{r_n} x_n) = \varepsilon^{\sigma} V(x_1, ..., x_n). \tag{3.41}$$

Let $x = (x_1, ..., x_n)^T$ and $f(x) = (f_1(x), ..., f_n(x))^T$ be a continuous vector field. $f(x)$ is said to be homogeneous of degree $\kappa \in R$ with respect to $(r_1, ..., r_n)$ if for any $\varepsilon > 0$,

$$f_i(\varepsilon^{r_1} x_1, ..., \varepsilon^{r_n} x_n) = \varepsilon^{\kappa + r_i} f_i(x), \quad i = 1, ..., n, \quad x \in R^n. \tag{3.42}$$

A given system $\dot{x} = f(x)$ is said to be homogeneous if $f(x)$ is homogeneous.

For a homogeneous system, we have the following lemma.

Lemma 3.10. *[67] Suppose $\dot{x} = f(x)$ is homogeneous of degree κ. Then the origin of the system is finite-time stable if the origin is asymptotically stable and $\kappa < 0$.*

Lemma 3.11. *[44] Consider the following system*

$$\dot{x} = f(x) + \tilde{f}(x), \quad x \in R^n, f(0) = 0, \tilde{f}(0) = 0 \tag{3.43}$$

where $f(x)$ is a continuous homogeneous vector filed of degree $\kappa < 0$ with respect to $(r_1, ..., r_n)$. Assume $x = 0$ is an asymptotical stable equilibrium of the system $\dot{x} = f(x)$. Then $x = 0$ is also a locally finite-time stable equilibrium of the system (3.43) if

$$\lim_{\varepsilon \to 0} \frac{\tilde{f}_i(\varepsilon^{r_1} x_1, ..., \varepsilon^{r_n} x_n)}{\varepsilon^{\kappa + r_i}} = 0, \quad i = 1, ..., n, \forall x \neq 0. \tag{3.44}$$

3.6.1 Proof of Fact 3.1

Consider a Lyapunov function $V_0 = \dfrac{1}{1 + k_1} |x|^{1 + k_1} + \dfrac{1}{2} y^2$. Then we get $\dot{V}_0 = -|y|^{1 + k_2} \leq 0$. Note that $\dot{V}_0 \equiv 0$ implies $y \equiv 0$, which further implies $x \equiv 0$. Thus direct application of LaSalle's invariant set theorem implies that the zero solution of (3.18) is asymptotically stable. Moreover, (3.18) is homogeneous of a negative degree $\kappa = k_2 - 1 < 0$ with respect to $(\frac{2}{1 + k_1}, 1)$, thus based on Lemma 3.10 we know (3.18) is globally finite-time stable.

4

Stabilization Control of Underactuated Ships with Input Saturation

In this chapter, stabilization control of underactuated ships in the presence of input saturation and disturbances is addressed. By introducing a virtual reference and proposing a novel saturated disturbance observer, the stabilization control of underactuated ships is transformed into a path-following problem, thus the error dynamics are divided into a cascade of two subsystems, and the torques in surge and yaw axis are designed separately with the backstepping technique. It is shown that the closed-loop system is stable and the position errors converge to the origin asymptotically. Simulation results also verify the effectiveness of the proposed scheme.

4.1 Introduction

The stabilization control of underactuated ships has received lots of attention over the past few decades. There has been significant interest in control of underactuated ships and various control schemes are proposed for such a problem. Normally the tracking and stabilization control of underactuated ships are investigated separately due to the existence of underactuation, and the control schemes cannot be applied to each other. In tracking control, backstepping and the Lyapunov function approach are used extensively. For example, in [17], two effective control schemes are presented to solve the global tracking control problem for underactuated ships with the Lyapunov method by exploiting the inherent cascade interconnected structure of the ship dynamics. In [18, 21] the tracking and path following control of underactuated ships are solved by backstepping techniques. In [22] a controller that achieves practical tracking of arbitrary reference trajectories for underactuated ships using the transverse function approach. However, the stabilization control of underactuated ships is normally solved by exploring the chained structure of the dynamics of the ship through variable exchange. In [82], the stabilization of underactuated ships is reduced to the stabilization of a third-order chained form by a change of coordinates. Subsequent related works on the control of underactuated ships include, but are not limited to, [45, 16, 46, 42, 20] and many references therein.

In practice, an underactuated ship is ultimately driven by a motor which can only provide limited amount of torque. Thus the magnitude of the control signal is always constrained. Under such a constraint, the controllability of the underactuated ships is seriously affected. Thus failing to tackle the input saturation constraint may worsen the control performance severely limited, and even give rise to undesirable inaccuracy or lead to instability. Therefore, the development of a control scheme for an underactuated ship with input saturation has been a difficult task with practical interest and theoretical significance. The tracking control of underactuated ships with input saturation have received attentions and several works have been proposed. For example, In [47] a tracking control method is presented to address such an issue by employing the linearization and dynamic surface control, only ensuring bounded tracking errors. In [83] a tracking control scheme is proposed for a underactuated ship with a reference under PE condition. In [84] a bounded tracking control scheme is proposed under a coordinate transformation. However, all these tracking control schemes cannot be applied to the stabilization of underactuated ships with input saturation. This is because through coordinates transformation, the parameters of the reference play an important role in the control design in the above references. Also these schemes cannot deal with the case when disturbances exist in the dynamics of the ship. There is no literature dealing with stabilization of underactuated ships with input saturation as far as we know, due to its difficulty.

In this chapter, we consider the stabilization control of underactuated ships with input saturation and external disturbances introduced by wave, wind or ocean currents. We solve this problem in the following steps. Firstly we propose a virtual reference, which lead to the origin and the parameters of the reference are free to design. Thus the stabilization of underactuated ships is transformed into a path following problem, where the error dynamic system can be divided into a cascade of two subsystems so that the torques in the surge and yaw axis can be designed separately based on the two subsystems. A new finite-time control scheme together with a new saturated finite-time disturbance observer are proposed to design the torque in yaw which enables the two subsystems to be de-coupled within finite time. After decoupling, a new backstepping scheme similar to that in [48] is proposed for the surge subsystem. The designed controllers for the two subsystems are analyzed, respectively. Then the overall system is studied. It is shown that the closed-loop system is stable and the stabilization errors converge to the origin asymptotically. The main contributions of this chapter are listed as follows: 1) A new virtual-path based approach is proposed for underactuated ships and the stabilization control is transformed to path-following control. 2) A new finite-time saturated disturbance observer is proposed for underactuated ships to estimate the external disturbances. To illustrate the effectiveness of the controller, simulation studies are conducted.

4.2 Problem Formation

4.2.1 Underactuated Ship Model

We consider the ship as shown in Figure 4.1. The only two propellers are the torques in the surge and yaw axis. The motion of the ship dynamics is described by the following differential equations

$$\dot{x} = u \cos \phi - v \sin \phi$$
$$\dot{y} = u \sin \phi + v \cos \phi$$
$$\dot{\phi} = r \tag{4.1}$$

$$\dot{u} = \frac{m_2}{m_1} vr - \frac{d_1}{m_1} u + \frac{1}{m_1} Sat_u(\tau_u) + d_u$$
$$\dot{v} = -\frac{m_1}{m_2} ur - \frac{d_2}{m_2} v + d_v$$
$$\dot{r} = \frac{m_1 - m_2}{m_3} uv - \frac{d_3}{m_3} r + \frac{1}{m_3} Sat_r(\tau_r) + d_r \tag{4.2}$$

where (x, y) denotes the position of the ship surface in X and Y directions, respectively, ϕ denotes the heading angle of the ship, u, v and r are the surge, sway and yaw velocities, respectively. τ_u and τ_r represent the torques in surge and yaw axis, respectively, which are considered as actual control inputs. Since there is no torque introduced in the sway direction, the system modeled by (4.1)-(4.2) is underactuated. The positive constants m_j denote the ship inertia including added mass, and d_j, $1 \leq j \leq 3$ represent the hydrodynamic damping coefficients. The saturation function is defined as follows

$$Sat_\star(x) = \begin{cases} x & \text{if } |x| < \tau_{\star \max} \\ \tau_{\star \max} & \text{if } |x| \geq \tau_{\star \max} \end{cases} \tag{4.3}$$

where $\star$ denotes u or r respectively and $\tau_{u \max}$ and $\tau_{r \max}$ are their respective saturation limits.

Our control objective is formally stated as follows.

• *Control Objective*: Design the control inputs τ_u and τ_r to force the ship, as modeled in (4.1)-(4.2) subject to input saturation, from initial position (x_0, y_0).

4.2.2 Virtual Reference and Variable Transformation

To facilitate the control design, firstly we define a path on the plane, which is denoted by $\Omega = (x_d, y_d)$ and generated by the following filter

$$\dot{x}_d = -2x_d + y_d$$
$$\dot{y}_d = -x_d - 2y_d \tag{4.4}$$

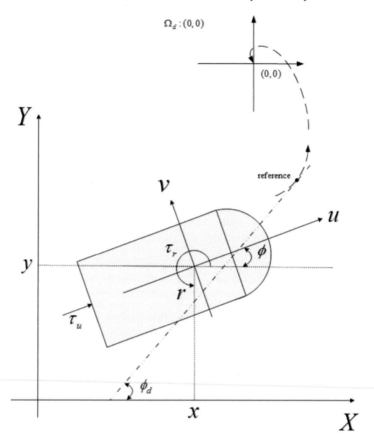

Fig. 4.1. The coordinates of an underactuated ship.

with $(x_d, y_d) = (x_0, y_0)$, which means

$$\begin{bmatrix} x_d(t) \\ y_d(t) \end{bmatrix} = \begin{bmatrix} x_0 e^{-2t} \cos(t) + y_0 e^{-2t} \sin(t) \\ y_0 e^{-2t} \cos(t) - x_0 e^{-2t} \sin(t) \end{bmatrix}. \tag{4.5}$$

Thus the control objective turns to design the control inputs τ_u and τ_r to force the ship to follow the reference $(x_d(s), y_d(s))$ given by (4.5) and shown in Fig.4.2., where s is a time-depending path variable to be designed later.

The desired surge and yaw speed with s are given as

$$u_d(s) = \sqrt{x_d'^2 + y_d'^2}\,\dot{s} := \bar{u}_d(s)\dot{s}$$

$$r_d(s) = \frac{x_d' y_d'' - x_d'' y_d'}{x_d'^2 + y_d'^2}\,\dot{s} := \bar{r}_d(s)\dot{s} \tag{4.6}$$

where $(\bullet)' = \partial(\bullet)/\partial s$, $(\bullet)'' = \partial^2(\bullet)/\partial s^2$. Define the errors $\{x_e, y_e, \phi_e\}$ as follows based on standard approach in control of underactuated ships

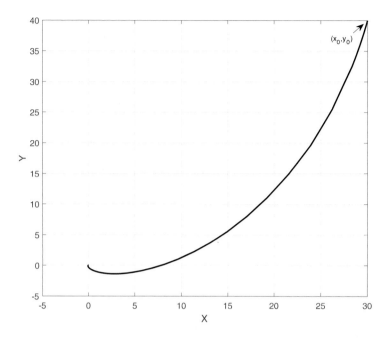

Fig. 4.2. Reference from initial position $(30, 40)$.

$$\begin{bmatrix} x_e \\ y_e \\ \phi_e \end{bmatrix} = \begin{bmatrix} \cos\phi & \sin\phi & 0 \\ -\sin\phi & \cos\phi & 0 \\ 0 & 0 & 1 \end{bmatrix} \begin{bmatrix} x - x_d \\ y - y_d \\ \phi - \phi_d \end{bmatrix} \tag{4.7}$$

where $\phi_d(s)$ is the angle between the path and the X-axis and is defined as $\phi_d(s) = \arctan(y_d'(s)/x_d'(s))$. Differentiating both sides of (4.7) along (4.1) results in the kinematic error dynamics:

$$\dot{x}_e = u - u_d \cos(\phi_e) + ry_e$$
$$\dot{y}_e = v + u_d \sin(\phi_e) - rx_e$$
$$\dot{\phi}_e = r - r_d. \tag{4.8}$$

$\dot{s}$ is defined as

$$\dot{s} = \frac{\bar{u}}{\sqrt{x_d^2 + y_d^2}} \tag{4.9}$$

where $\bar{u}$ is defined as

$$\bar{u} = \begin{cases} u_0, & \text{if } \sqrt{x_d^2 + y_d^2} \geq \bar{a} \\ \beta\sqrt{x_d^2 + y_d^2}, & \text{if } 0 \leq \sqrt{x_d^2 + y_d^2} \leq \bar{a} - \beta_1\bar{r} \\ u_0 - \bar{r} + \sqrt{\bar{r}^2 - (\sqrt{x_d^2 + y_d^2} - \bar{a})^2}, & \\ & \text{if } \bar{a} - \beta_1\bar{r} < \sqrt{x_d^2 + y_d^2} < \bar{a} \end{cases} \tag{4.10}$$

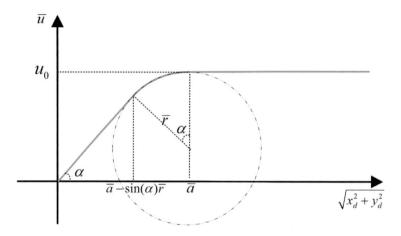

Fig. 4.3. $\bar{u}$ with respect to $\sqrt{x_d^2 + y_d^2}$.

where u_0, α and $\bar{r}$ are parameters to be designed, $\bar{a} = (u_0-\bar{r})\cot(\alpha)+\bar{r}/\sin(\alpha)$ is positive constant, $\beta = \tan(\alpha)$ and $\beta_1 = \sin(\alpha)$ and is shown in Fig.4.3.

Remark 4.1. The designed virtual reference with design parameter variable $\dot{s}$ has the following features which will benefit the control design:
(a) First the virtual reference will move with an adjustable constant surge speed u_d then with a constant yaw speed r_d.
(b) It could be checked that $\bar{u}_d = \sqrt{5(x_d^2 + y_d^2)}$, $\bar{r}_d = -1$, $|r_d| < \beta$, $|\dot{u}_d| \leq \sqrt{5}\beta(\bar{a} - \beta_1\bar{r})$, $|\dot{r}_d| < \frac{2u_0^2}{\bar{a}^2}$; thus u_d, r_d, $\dot{u}_d$ and $\dot{r}_d$ could also be arbitrarily small by design parameter.
(c) The desired yaw speed satisfies $r_d > 0$, thus the PE condition of r_d is satisfied, which will simplify the control design greatly.

To achieve the control objective, the following assumptions are imposed.

Assumption 4.1 *The disturbance d_u, d_v and d_r are bounded. Furthermore, there exists positive constants ϵ_{r1} and ϵ_{r2} such that $|d_r| < \epsilon_{r1}$, $|\dot{d}_r| < \epsilon_{r2}$.*

Assumption 4.2 *There exists a positive constant $\Delta > 2m_3\epsilon_{r1}$ such that*

$$\frac{|m_1 - m_2|}{2d_1\alpha}\tau_{u\,\max}^2 + \Delta \leq \tau_{r\,\max} \tag{4.11}$$

where $\alpha = \min\{\frac{d_1}{m_1}, \frac{2d_2}{m_2}\}$.

Remark 4.2. The main difficulty of stabilization control of underactuated ships, as well as the major difference between the tracking and stabilization control of underactuated ships, is that stabilization control could not make use of the variable transformation (4.7) directly. Although in [19] a stabilization of an underactuated ship based on backtepping is proposed, however, the 'dimension explosion' of the backstepping makes it impossible to be extended to input saturation control. Thus in this paper, through the definition of the "reference signal", the stabilization control is transformed into a path-following control of an underactuated ship, so we can make use of the variable transformation (4.7) to divide the original kinematic model and dynamic model of an underactuated ship into a cascade of two subsystems, i.e., (ϕ_e, r) subsystem and (x_e, y_e, v, u) subsystem.

Remark 4.3. We assume the wind, wave or ocean current are at acceptable levels such that the bounds of the disturbance at the sway direct are known. Assumption 4.2 is the restrictions for the saturation limits and the reference that is needed for controller design.

4.3 Controller Design

The control design procedure can be summarized as follows. Firstly (4.8) will be divided into a cascade of two subsystems, i.e., (ϕ_e, r) subsystem and (x_e, y_e, v, u) subsystem. The methodology of this paper is as follows: Control τ_r is designed such that ϕ_e converge to 0 within finite time T, while it guarantees that $|\tau_r| < \tau_{r\max}$. For $t > T$ the two subsystems are de-coupled since the interaction effects are only from (ϕ_e, r) subsystem to (x_e, y_e, v, u) subsystem, i.e., a one-way interaction. Control τ_u is designed based on (x_e, y_e, v, u) subsystem with a new backstepping technique in such a way that τ_u is the output of a filter. It is then shown that the overall closed-loop system is stable and the position errors converge to the origin asymptotically.

Firstly we propose a saturated finite-time disturbance observer for a dynamic system $\dot{x} = u + d$ where d is a disturbance with $|d| \leq \mathcal{D}$ and $|\dot{d}| \leq \mathcal{L}$.

Lemma 4.4. *The following control input and saturated disturbance observer guarantee $\rho = -d$ within finite time.*

$$u = -kSat_{\varepsilon_1}(sig^{\frac{1}{2}}(x)) + Sat_{\varepsilon_2}(\rho)$$

$$\dot{\rho} = -\frac{k}{2}sgn(x) \tag{4.12}$$

where $k = 1 + m$ with m being a positive constant such that $m \geq 2\mathcal{L}^2 + \frac{1}{2}$, ε_1 is a positive constant such that $k\varepsilon_1 = \mathcal{D} + \epsilon_1$ with ϵ_1 being an arbitrarily small positive constant, and ε_2 is chosen as $\varepsilon_2 = \mathcal{D} + 2\varepsilon_1$.

Proof: Taking (4.12) into the dynamic system yields

$$\dot{x} = -kSat_{\varepsilon_1}(sig^{\frac{1}{2}}(x)) + Sat_{\varepsilon_2}(\rho) + d$$

$$\dot{\rho} = -\frac{k}{2}sgn(x). \tag{4.13}$$

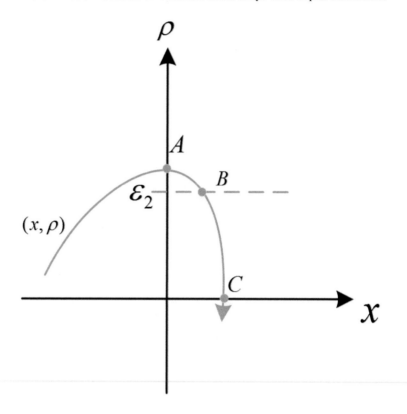

Fig. 4.4. Geometrical trajectory of (x, ρ).

Define a Lyapunov function $V = k|x| + \int_0^\rho Sat_{\varepsilon_2}(\tau)d\tau$, then

$$\dot{V} = -k\left(kSat_{\varepsilon_1}(|x|^{\frac{1}{2}}) - d\right) \tag{4.14}$$

which means $\dot{V} \leq -\epsilon_1$ when $|x|^{\frac{1}{2}} \geq \varepsilon_1$, then $Sat_{\varepsilon_1}(\cdot)$ for $sig^{\frac{1}{2}}(x)$ will not take effect within finite-time. It yields

$$\dot{x} = -ksig^{\frac{1}{2}}(x) + Sat_{\varepsilon_2}(\rho) + d$$
$$\dot{\rho} = -\frac{k}{2}sgn(x). \tag{4.15}$$

Now based on (4.15) it is easy to analyze geometrically from Fig.4.4 that (x, ρ) will reach point A, and then point B and point C within finite time. Suppose at (x, ρ) reach point C at $t = T$. Clearly at point C the $Sat_{\varepsilon_2}(\cdot)$ for ρ does not take effect. Firstly apply a coordinates transformation as $(sig^{\frac{1}{2}}(x), \rho) = (z_1, z_2)$, then

$$\dot{z}_1 = \frac{1}{2}|z_1|^{-1}[-kz_1 + z_2 + d]$$

$$\dot{z}_2 = -\frac{k}{2}sgn(z_1). \tag{4.16}$$

Make define a new variable as $\bar{z}_2 = z_2 + d - z_1$, then

$$\dot{z}_1 = \frac{1}{2}|z_1|^{-1}[-mz_1 + \bar{z}_2]$$

$$\dot{\bar{z}}_2 = -\frac{k}{2}sgn(z_1) + \dot{d} - \frac{1}{2}|z_1|^{-1}[-mz_1 + \bar{z}_2]. \tag{4.17}$$

Define a Lyapunov function $V = \frac{1}{2}(z_1^2 + \bar{z}_2^2)$, then

$$\dot{V} = -\frac{1}{2}|z_1|^{-1}(mz_1^2 - 2\dot{d}|z_1|\bar{z}_2 + \bar{z}_2^2) \leq -\frac{1}{2}|z_1|^{-1}V, \tag{4.18}$$

thus $|\bar{z}_2| \leq \sqrt{2V(T)} \leq \varepsilon_1$, which means $|z_2| \leq |\bar{z}_2| + \mathcal{D} + |z_1| = \varepsilon_2$. Therefore $Sat_{\varepsilon_2}(\cdot)$ does not take effect for $t \geq T$ and (4.18) holds for $t \geq T$. As $|z_1| \leq \sqrt{2}V^{\frac{1}{2}}$; then $|z_1| \geq \sqrt{1/2}V^{-\frac{1}{2}}$; thus

$$\dot{V} \leq -\frac{1}{2\sqrt{2}}V^{\frac{1}{2}}, \tag{4.19}$$

thus z_1 and z_2 are finite time stable. This completes the proof. □

Now we develop the following intermediate result to reveal the bounds of u, v and r.

Lemma 4.5. *The variables u, v and r are bounded and their respective bounds satisfy*

$$\sup_{\forall t>0}|u| \leq \sqrt{\frac{m_2}{2m_1 d_1 \alpha}}\tau_{u\,max}, \quad \sup_{\forall t>0}|v| \leq \sqrt{\frac{m_1}{2m_2 d_1 \alpha}}\tau_{u\,max},$$

$$\sup_{\forall t>0}|r| \leq \frac{\tau_{u\,max}}{\sqrt{m_3\beta d_1}} + \frac{\tau_{r\,max}}{\sqrt{m_3\beta d_3}} \tag{4.20}$$

where $\beta = \min\{\frac{d_1}{m_1}, \frac{2d_2}{m_2}, \frac{d_3}{m_3}\}$.

Proof: Consider a Lyapunov function $V_0 = \frac{m_1}{2m_2}u^2 + \frac{m_2}{2m_1}v^2$, whose derivative is

$$\dot{V}_0 = -\frac{d_1}{m_2}u^2 - \frac{d_2}{m_1}v^2 + \frac{1}{m_2}uSat_u(\tau_u)$$

$$\leq -\alpha V_0 + \frac{1}{2d_1}\tau_{u\,max}^2.$$

Without loss of generality, let $u(0) = 0$ and $v(0) = 0$. Thus $V_0 \leq \frac{\tau_{u\,max}^2}{2d_1\alpha}$. So $|u| \leq \sqrt{\frac{m_2}{2m_1 d_1\alpha}}\tau_{u\,max}$ and $|v| \leq \sqrt{\frac{m_1}{2m_2 d_1\alpha}}\tau_{u\,max}$. For r, consider $V = \frac{m_1}{2}u^2 + \frac{m_2}{2}v^2 + \frac{m_3}{2}r^2$, whose derivative is

$$\dot{V} = -d_1 u^2 - d_2 v^2 - d_3 r^2 + u Sat_u(\tau_u) + r Sat_r(\tau_r)$$

$$\leq -\beta V + \frac{\tau_{u\,\max}^2}{2d_1} + \frac{\tau_{r\,\max}^2}{2d_3},$$

so $V \leq \frac{\tau_{u\,\max}^2}{2\beta d_1} + \frac{\tau_{r\,\max}^2}{2\beta d_3}$ and $|r| \leq \sqrt{\frac{\tau_{u\,\max}^2}{m_3 \beta d_1} + \frac{\tau_{r\,\max}^2}{m_3 \beta d_3}} \leq \frac{\tau_{u\,\max}}{\sqrt{m_3 \beta d_1}} + \frac{\tau_{r\,\max}}{\sqrt{m_3 \beta d_3}}$. This completes the proof. □

Now we design τ_r with the aim that $\sup |\tau_r| \leq \tau_{r\,\max}$ and ϕ_e defined in (4.8) converges to zero within finite time. Consider the following system

$$\dot{\phi}_e = r_e$$

$$\dot{r}_e = \frac{m_1 - m_2}{m_3} uv - \frac{d_3}{m_3} r + \frac{1}{m_3} Sat_r(\tau_r) - \dot{r}_d + d_r \qquad (4.21)$$

where $r_e = r - r_d$. First of all, define $sig(x)^k = sign(x)|x|^k$, then τ_r is designed as

$$\tau_r = (m_2 - m_1)uv + m_3 \dot{r}_d + d_3 r_d$$
$$- m_3 Sat_{\varepsilon_1}\left(sig^{k_1}(\phi_e) \right) - m_3 Sat_{\varepsilon_2}\left(sig^{k_2}(r_e) \right)$$
$$- k_3 m_3 Sat_{\varepsilon_3}(sig^{\frac{1}{2}}(w)) + m_3 Sat_{\varepsilon_4}\left(\varrho \right)$$

$$\dot{\varrho} = -\frac{k_3}{2} sgn(s) \qquad (4.22)$$

where ε_1, ε_2 and ε_3 are positive design parameters, k_1 and k_2 are two positive constants satisfying $0 < k_1 < 1$, $k_2 = \frac{2k_1}{1 + k_1}$, k_3 is a positive constant such that $k_3 > \epsilon_{r2}$. ε_3 is chosen such that $k_3 > 2\epsilon_{r2}^2 + 1.5$, $\varepsilon_3 = \epsilon_{r1}/\varepsilon_3$, $\varepsilon_4 = \epsilon_{r1} + 2\varepsilon_3$. w is a sliding variable defined as

$$w = r_e - r_e(0) - \int_0^t \tau_{r0}(\iota)d\iota \qquad (4.23)$$

where $\tau_{r0} = (m_2 - m_1)uv + m_3 \dot{r}_d + d_3 r_d - m_3 Sat_{\varepsilon_1}(sig^{k_1}(\phi_e)) - m_3 Sat_{\varepsilon_2}(sig^{k_2}(r_e))$. The time-derivative of s is

$$\dot{w} = -k_3 Sat_{\varepsilon_3}(sig^{\frac{1}{2}}(w)) + Sat_{\varepsilon_4}(\varrho) + d_r$$

$$\dot{\varrho} = -\frac{k_3}{2} sgn(w). \qquad (4.24)$$

Thus based on Lemma 4.4, w and ϱ will be stabilized within finite time, thus substitute (4.22) into (4.21),

$$\dot{\phi}_e = r_e$$

$$\dot{r}_e = -Sat_{\varepsilon_1}\left(sig^{k_1}(\phi_e) \right) - Sat_{\varepsilon_2}\left(sig^{k_2}(r_e) \right) - \frac{d_3}{m_3} r_e. \qquad (4.25)$$

Then the following lemma is established.

Lemma 4.6. *With controller (4.22), ϕ_e and r_e in (4.25) will converge to zero within finite time. Furthermore, it is ensured that $|\tau_r| \leq \tau_{r\,\max}\ \forall t$ by choosing appropriate design parameters ε_1 and ε_2.*

Proof: Consider the following Lyapunov function

$$V_1 = \int_0^{\phi_e} Sat_{\varepsilon_1}(sig(\iota)^{k_1})d\iota + \frac{1}{2}r_e^2, \tag{4.26}$$

whose derivative along (4.25) is

$$\dot{V}_1 = -\frac{d_3}{m_3}r_e^2 - r_e sat_{\varepsilon_2}(sig(r_e)^{k_2}). \tag{4.27}$$

From (4.27), r_e is globally asymptotically convergent. Similarly it can also be proved that ϕ_e is globally asymptotically convergent. Note that the saturation in (4.25) will not take effect when

$$|\phi_e| \leq \varepsilon_1^{1/k_1}, \quad |r_e| \leq \varepsilon_2^{1/k_2}. \tag{4.28}$$

Thus we define a set

$$\Pi = \{(\phi_e, r_e) : \frac{1}{1+k_1}|\phi_e|^{1+k_1} + \frac{1}{2}r_e^2 \leq \theta\} \tag{4.29}$$

where $\theta = \min\{\frac{\varepsilon_1^{(1+k_1)/k_1}}{1+k_1}, \frac{1}{2}\varepsilon_2^{2/k_2}\}$. Now consider $V_2 = \frac{1}{1+k_1}|\phi_e|^{1+k_1} + \frac{1}{2}r_e^2$. When $\phi_e \in \Pi$ and $r_e \in \Pi$, the saturation does not take effect, and it can be easily checked that $\dot{V}_2 = -\frac{d_3}{m_3}r_e^2 - |r_e|^{1+k_2} \leq 0$, thus Π is an invariant set.

When $(\phi_e, r_e) \notin \Pi$, (ϕ_e, r_e) approaches Π within finite time from (4.27), and after (ϕ_e, r_e) reaches Π, we have

$$\dot{\phi}_e = r_e$$
$$\dot{r}_e = -\frac{d_3}{m_3}r_e - \left(sig(\phi_e)^{k_1} + sig(r_e)^{k_2}\right). \tag{4.30}$$

Next we show that (4.30) is finite-time stable. Based on Theorem 1 in [43], we know that (4.30) is finite-time stable if $(\phi_e, r_e) \in \Pi$.

So in summary, if $(\phi_e(0), r_e(0)) \notin \Pi$, (ϕ_e, r_e) will reach Π within finite time, and when $(\phi_e, r_e) \in \Pi$, (ϕ_e, r_e) will converge to the origin within finite time. Thus (ϕ_e, r_e) will approach to zero within a total of finite time T and therefore(4.25) is globally finite time stable.

Also from (4.22) and Lemma 4.4,

$$|\tau_r| \leq |m_2 - m_1| \frac{1}{2d_1\alpha} \tau_{u\,max}^2 + m_3(\varepsilon_1 + \varepsilon_2 + k_3\varepsilon_3 + \varepsilon_4) + \left|m_3\dot{r}_d + d_3r_d\right|.$$

Based on Assumption 4.1, choose ε_1, ε_2, ε_3 and ε_4 to satisfy

$$\left|m_3\dot{r}_d + d_3r_d\right| + m_3(\varepsilon_1 + \varepsilon_2 + k_3\varepsilon_3 + \varepsilon_4) \leq \Delta, \tag{4.31}$$

then we have

$$|\tau_r| \leq \tau_{r\,max}, \tag{4.32}$$

which means the saturation does not take effect for the yaw subsystem. This completes the proof.

Since ϕ_e will converge to α_{ϕ_e} within a finite time T, and due to the saturation, u, v and r are bounded; thus for $t \leq T$, x_e, y_e, ϕ_e, u and r are bounded. As for $t > T$, $\phi_e = 0$, and then the (x_e, y_e, v, u) subsystem becomes

$$\dot{y}_e = v - r_d x_e$$
$$\dot{x}_e = u - u_d + r_d y_e$$
$$\dot{v} = -\frac{m_1}{m_2}ur - \frac{d_2}{m_2}v + d_v$$
$$\dot{u} = \frac{m_2}{m_1}vr - \frac{d_1}{m_1}u + \frac{1}{m_1}Sat_u(\tau_u) + d_u. \tag{4.33}$$

The model in (4.33) enables us to design τ_u for $t > T$ without considering the coupling effects from (ϕ_e, r_e) subsystem. Due to the one-way interaction, the designed control τ_u can still be applied from the initial time $t = 0$. As $Sat_u(\cdot)$ is non-smooth, thus backstepping technique, see [8], cannot be applied here. To overcome this difficulty, the saturation will be approximated by a smooth function defined as in [48]

$$s(\tau_u) = \tau_{u\,max} \tanh(\frac{\tau_u}{\tau_{u\,max}}) = \tau_{u\,max} \frac{e^{\tau_u/\tau_{u\,max}} - e^{-\tau_u/\tau_{u\,max}}}{e^{\tau_u/\tau_{u\,max}} + e^{-\tau_u/\tau_{u\,max}}} \tag{4.34}$$

Then $Sat_u(\cdot)$ can be expressed as

$$Sat_u(\tau_u) = s(\tau_u) + d(t) \tag{4.35}$$

where $d(t) = Sat_u(\tau_u) - s(\tau_u)$ is a bounded function which can be deemed as a 'disturbance' bounded as

$$|d(t)| = |Sat_u(\tau_u) - s(\tau_u)| \leq \tau_{u\,max}(1 - \tanh(1)). \tag{4.36}$$

Figure 4.5 shows an example of approximating the saturation function. To design the control τ_u, we augment the system (4.33) by including the saturation approximation function and the resulting approximation error as well as an

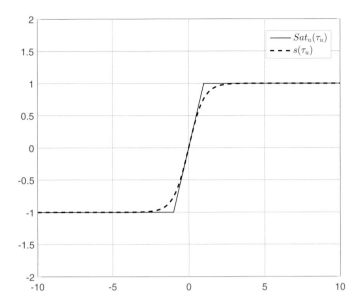

Fig. 4.5. Approximation of the saturation function $Sat_1(\tau_u)$.

additional user chosen first-order filter as follows

$$\dot{y}_e = v - r_d x_e$$
$$\dot{x}_e = u - u_d + r_d y_e$$
$$\dot{v} = -\frac{m_1}{m_2} u r_d - \frac{d_2}{m_2} v + d_v$$
$$\dot{u} = \frac{m_2}{m_1} v r_d - \frac{d_1}{m_1} u + \frac{1}{m_1} s(\tau_u) + \frac{1}{m_1} d(t) + d_u$$
$$\dot{\tau}_u = -c\tau_u + \alpha \qquad\qquad (4.37)$$

where c is a positive design parameter and α is an auxiliary intermediate control to be designed in the backstepping approach. Let α_{y_e} and α_{x_e} be the virtual controls of x_e and u, and α_{τ_u} be the virtual control of $s(\tau_u)$. Then following the standard backstepping design steps [8], these virtual controls and the intermediate control α in (4.37) can be obtained as summarized in Table 4.1 where $N(\chi)$ is a Nassbaum function, $\gamma = 1 + \frac{m_1}{m_2}$, D_1 is the bound of $\frac{d_v}{r_d}$, D_2 is the bound of $d(t) + m_1 d_u + \frac{d_2}{m_2 r_d} d_v$, D_3 is the bound of $\frac{d_2^2}{m_2^2 r_d} d_v$, and $\hat{D}_i$ are the estimation of D_i, $i = 1, 2, 3$ respective, and $\tilde{D}_i = \hat{D}_i - D_i$.

Note that the last equation of (4.37) is not part of the ship model. It is actually an user designed filter that generates the final control τ_u with the designed α in (4.37) as its input. The controller for subsystem $(\tilde{x}_e, y_e, \tilde{u}_e, \tilde{\tau}_u)$

Virtual control errors:

$$\tilde{x}_e = x_e - \alpha_{y_e}$$

$$\tilde{u} = u - \alpha_{x_e}$$

$$\tilde{\tau}_u = \alpha_{\tau_u} - s(\tau_u) \tag{4.38}$$

Virtual Controllers and Intermediate Controller:

$$\alpha_{y_e} = c_1 r_d y_e + v/r_d$$

$$\alpha_{x_e} = -c_2 \tilde{x}_e + \frac{1}{\gamma}\left[u_d - \frac{\partial \alpha_{y_e}}{\partial r_d} \dot{r}_d - \frac{\partial \alpha_{y_e}}{\partial y_e} \dot{y}_e \right.$$

$$\left. + \frac{d_2}{m_2 r_d} v - \frac{\tilde{x}_e}{\sqrt{\tilde{x}_e^2 + \eta^2}} \hat{D}_1 \right]$$

$$\alpha_{\tau_u} = d_1 \alpha_{x_e} - m_2 v r_d - m_1\left(\frac{\partial \alpha_{x_e}}{\partial \tilde{x}_e} \dot{\tilde{x}}_e + \frac{\partial \alpha_{x_e}}{\partial y_e} \dot{y}_e \right.$$

$$+ \frac{\partial \alpha_{x_e}}{\partial r_d} \dot{r}_d + \frac{\partial \alpha_{x_e}}{\partial v}\left(\frac{m_1}{m_2} \alpha_{x_e} r_d - \frac{d_2}{m_2} v \right) + \frac{\partial \alpha_{x_e}}{\partial \eta} \dot{\eta}$$

$$\left. + \frac{\partial \alpha_{x_e}}{\partial u_d} \dot{u}_d + \frac{\partial \alpha_{x_e}}{\partial \hat{D}_1} \dot{\hat{D}}_1 \right) - \frac{\tilde{u}}{\sqrt{\tilde{u}^2 + \eta^2}} \hat{D}_2$$

$$\alpha = N(\chi)\bar{\alpha}$$

$$\bar{\alpha} = -c_3 \tilde{\tau}_u + c\tau_u \frac{\partial s(\tau_u)}{\partial \tau_u} + \frac{\partial \alpha_{\tau_u}}{\partial x_e} \dot{x}_e$$

$$+ \frac{\partial \alpha_{\tau_u}}{\partial y_e} \dot{y}_e + \frac{\partial \alpha_{\tau_u}}{\partial r_d} \dot{r}_d + \frac{\partial \alpha_{\tau_u}}{\partial \hat{D}_1} \dot{\hat{D}}_1 + \frac{\partial \alpha_{\tau_u}}{\partial \hat{D}_2} \dot{\hat{D}}_2 +$$

$$\frac{\partial \alpha_{\tau_u}}{\partial v}\left(-\frac{m_1}{m_2} u r_d - \frac{d_2}{m_2} v \right) + \frac{\partial \alpha_{\tau_u}}{\partial u_d} \dot{u}_d - \frac{\tilde{\tau}_u}{\sqrt{\tilde{\tau}_u^2 + \eta^2}} \hat{D}_3$$

$$N(\chi) = \chi^2 \cos(\frac{\pi \chi}{2}), \dot{\chi} = \bar{\alpha} \tilde{\tau}_u$$

$$\dot{\hat{D}}_1 = \frac{c_4 \tilde{x}_e^2}{\sqrt{\tilde{x}_e^2 + \eta^2}}, \quad \dot{\hat{D}}_2 = \frac{c_5 \tilde{u}^2}{\sqrt{\tilde{u}^2 + \eta^2}}, \quad \dot{\hat{D}}_3 = \frac{c_6 \tilde{\tau}_u^2}{\sqrt{\tilde{\tau}_u^2 + \eta^2}} \tag{4.39}$$

Table 4.1. Summary of virtual control errors, virtual controls and intermediate control.

consists of (4.37)-(4.39). Note that the following inequality

$$0 \le |x| - \frac{x^2}{\sqrt{x^2 + \eta^2}} \le \eta \tag{4.40}$$

holds for $x \in \Re$ and $\eta > 0$;

Consider the following Lyapunov function

$$V_3 = \frac{1}{2} y_e^2 + \frac{1}{2} \tilde{x}_e^2 + \frac{1}{2} \tilde{u}^2 + \frac{1}{2} \tilde{\tau}_u^2 + \frac{1}{2c_4} \tilde{D}_1^2 + \frac{1}{2c_5} \tilde{D}_2^2 + \frac{1}{2c_6} \tilde{D}_3^2, \tag{4.41}$$

and then we have

$$\dot{V}_3 \le (\xi N(\chi) - 1)\bar{a}\tilde{\tau}_u - \check{c}V_e + (D_1 + D_2 + D_3)\eta \qquad (4.42)$$

$$\le (\xi N(\chi) - 1)\bar{a}\tilde{\tau}_u - \check{c}V_3 + (D_1 + D_2 + D_3)\eta + \Theta \qquad (4.43)$$

where $V_e = \frac{1}{2}y_e^2 + \frac{1}{2}\tilde{x}_e^2 + \frac{1}{2}\tilde{u}^2 + \frac{1}{2}\tilde{\tau}_u^2$, $\check{c} = 2 \times \min\{c_1 r_d^2, c_2\gamma^2, d_1 + \frac{m_1}{m_1+m_2}(\frac{\dot{r}_d}{r_d} + d_2), c_3\} > 0$, $\xi = \frac{\partial s(\tau_u)}{\partial \tau_u} = \frac{4}{(e^{\tau_u/\tau_u \max} + e^{-\tau_u/\tau_u \max})^2} \in (0,1)$, $\Theta = \check{c}\max\{\frac{1}{2c_4}\tilde{D}_1^2 + \frac{1}{2c_5}\tilde{D}_2^2 + \frac{1}{2c_6}\tilde{D}_3^2\}$. With the proposed control, the following lemma is established.

Lemma 4.7. *Consider system (4.37) under the control of (4.37)-(4.39). If Assumption 4.1 and Assumption 4.2 are satisfied and $V_3(T)$ is bounded, then all signals in (4.37)-(4.39) are bounded for $t > T$.*

Proof: From (4.43) we have

$$V_3(t) \le V_3(T)e^{-\check{c}(t-T)} + \tilde{D}e^{-\check{c}t}\int_T^t \eta e^{\check{c}\sigma}\,d\sigma + e^{-\check{c}t}\int_T^t \Theta\,d\sigma$$

$$+ e^{-\check{c}t}\int_T^t (\xi N(\chi)\dot{\chi} - \dot{\chi})e^{\check{c}\sigma}\,d\sigma \qquad (4.44)$$

for $t > T$, where $\tilde{D} = D_1 + D_2 + D_3$. By choosing $\eta = e^{-\check{c}t}$, then $\tilde{D}e^{-\check{c}t}\int_T^t \eta e^{\check{c}\sigma}\,d\sigma$ is bounded. The boundedness of $e^{-\check{c}t}\int_T^t (\xi N(\chi)\dot{\chi} - \dot{\chi})e^{\check{c}\sigma}\,d\sigma$ can be established based on the Nussbaum properties via a contradiction argument. First define V_N on the time-period (t_i, t_j) as

$$V_N(t_i, t_j) = \int_{t_i}^{t_j} (\xi N(\chi) - 1)\dot{\chi}e^{-\check{c}(t_j - \sigma)}\,d\sigma. \qquad (4.45)$$

For notation convenience, let $V_N(t_i, t_j) = V_N(\chi_i, \chi_j)$, $T < t_i < t_j$. By noting the fact that $0 < \xi < 1$ and $0 < e^{-\check{c}(t_j - \sigma)} < 1$, we have

$$|V_N(\chi_i, \chi_j)| \le \int_{\chi_i}^{\chi_j} (|N(\chi)| + 1)\,d\chi$$

$$\le (\chi_j - \chi_i)\Big(sup_{\chi \in [\chi_i, \chi_j]}|N(\chi)| + 1\Big). \qquad (4.46)$$

For the Nussbaum function $N(\chi) = \chi^2 \cos(\frac{\pi\chi}{2})$, we know it is positive for $\chi \in (4m - 1, 4m + 1)$ and negative for $\chi \in (4m + 1, 4m + 3)$ with an integer m. Thus we consider two time periods: $[\chi_0, \chi_1] = [\chi_0, 4m + 1]$ and $[\chi_1, \chi_2] = [4m+1, 4m+3]$ for $\chi_0 > 0$ and m being a positive integer such that for $t \to t_f$, where t_f is an arbitrary positive constant, we have $\chi = \chi_2$. From (4.46) we know for $[\chi_0, \chi_1] = [\chi_0, 4m + 1]$, we have

$$V_N(\chi_0, \chi_2) = V_N(\chi_0, \chi_1) + V_N(\chi_1, \chi_2)$$

$$\le (4m + 1)^2\Big(-l_2((8m + 2)(1 - \Psi) + (1 - \Psi)^2)$$

$$+ 4m + 1 - \chi_0 + \frac{l_1 + l + 3}{(4m + 1)^2}\Big) \qquad (4.47)$$

where $l_2 = 2\Psi e^{-\tilde{c}*(t_2-t_1)} \cos(\pi\Psi/2) > 0$, $\Psi \in (0,1)$ and $l_3 = 2\Psi e^{-\tilde{c}*(t_2-t_1)} > 0$.

We now prove χ is bounded over $(T, t_f]$ by seeking a contradiction. Suppose χ is unbounded over $(T, t_f]$. Two cases should be considered: (i) χ has no upper-bound and (ii) χ has no lower-bound. If $\chi(t)$ has no upper bound on $(T, t_f]$. In this case, there must exist a time interval $[t_s, t_f]$ such that $\chi(t_s) > 0$, when $t \to t_f$, $\chi(t) = \infty$. However, from (4.47) we know $V_N(\chi_0, \chi_2) \to -\infty$ as $m \to \infty$. On the other hand, $V_3(t) > 0$ for all t. Thus we can always find a subsequence of the interval that leads to a contradiction. So $\chi(t)$ has an upper bound. On the other side, if $\chi(t)$ has no lower bound on $(T, t_f]$; define $\chi = -\lambda$; then λ has no upper bound accordingly. Since $N(\chi)$ is an even function, thus (4.43) becomes

$$
\begin{aligned}
V_3(t) &\leq V_3(T)e^{-\tilde{c}(t-T)} + \tilde{D}e^{-\tilde{c}t}\int_T^t \eta e^{\tilde{c}\sigma}d\sigma + e^{-\tilde{c}t}\int_T^t \Theta d\sigma \\
&\quad - \int_T^t (\xi N(\lambda) - 1)\dot{\lambda}e^{-\tilde{c}(t-\sigma)}d\sigma \\
&= V_3(T)e^{-\tilde{c}(t-T)} + \tilde{D}e^{-\tilde{c}t}\int_T^t \eta e^{\tilde{c}\sigma}d\sigma + e^{-\tilde{c}t}\int_T^t \Theta d\sigma - V_N(\lambda(T), \lambda(t)).
\end{aligned}
$$
(4.48)

Therefore there exists a time interval $[t_s, t_f]$ such that $\lambda(t)$ is monotonically increasing and $\lim_{t \to t_f} \lambda(t) = \infty$ with $\lambda(t_s) > 0$. Following the same procedure, we can construct a subsequence of the interval that leads to a contradiction. Therefore χ has a lower bound on $[T, t_f]$. Then from (4.46) V_N is bounded. The above argument holds for all $t_f > T$. So χ is bounded. Thus V_3 is also bounded, which means $\tilde{x}_e$, y_e, $\tilde{u}$ and $\tilde{\tau}_u$ are bounded. This completes the proof.

With Lemma 4.6 and Lemma 4.7, the following theorem is established.

Theorem 4.8. *Consider the closed loop system consisting of the underactuated ship dynamic systems (4.1)-(4.2) and the proposed controllers (4.22), (4.37) and (4.38)-(4.39) designed under Assumption 4.1 and Assumption 4.2. The orientational error ϕ_e converges to origin within finite time. The position errors $x_e(t)$ and $y_e(t)$ converge to the origin asymptotically.*

Proof: We consider $t \in (0, T)$. Since τ_u and τ_r are saturated, it is easy to be checked that u, v, r, x, y, ϕ, $\tilde{u}$, r_e, $\tilde{x}_e$, y_e, ϕ_e and their derivatives are bounded for $t \in (0, T)$. From (4.39) α_{τ_u} and $\dot{\alpha}_{\tau_u}$ are bounded. Thus from (4.38) we know $\tilde{\tau}_u$ is bounded. Then based on (4.37) and (4.39), we know τ_u and χ are bounded for $t \in (0, T)$. Thus from Lemma 4.6 and Lemma 4.7 we know the closed-loop system is stable and the orientational error ϕ_e converges to origin within finite time T.

From (4.42),

$$\lim_{T_1 \to \infty} \int_T^{T_1} V_e(t)dt \le \frac{V_3(T_1) - V_3(T)}{\tilde{c}}$$

$$+ \lim_{T_1 \to \infty} \frac{1}{\tilde{c}} \int_{\chi(T)}^{\chi(T_1)} (\xi N(\chi) - 1)d\chi + \lim_{T_1 \to \infty} \frac{1}{\tilde{c}} \int_T^{T_1} \tilde{D} e^{-\tilde{c}t} dt.$$

(4.49)

Since $\lim_{T_1 \to \infty} \int_{\chi(T)}^{\chi(T_1)} (\xi N(\chi) - 1)d\chi$ and $\lim_{T_1 \to \infty} \frac{1}{\tilde{c}} \int_T^{T_1} \tilde{D} e^{-\tilde{c}t} dt$ are bounded, thus $\lim_{T_1 \to \infty} \int_0^{T_1} V_e(t)dt$ is also bounded. Meanwhile it is obvious that $V_e(t)$ is uniformly continuous because its time derivative is bounded. Hence from Barbalat's lemma we have

$$\lim_{t \to \infty} V_e(t) = 0. \tag{4.50}$$

This completes the proof.

Remark 4.9. Our methodology is not only limited to the underactuated ships, but also applicable to the unsolved adaptive output feedback control problem of many other uncertain mechanical systems with input saturation, such as nonholonomic mobile robots, underwater vehicles, 4-DOF underactuated cranes etc.

4.4 Simulations

In this section, we carry out simulations to demonstrate the effectiveness of our controllers and to validate our constructive methodology for underactuated ships. The parameters of the underactuated ship from [47] are used, namely $m_1 = 120 \times 10^3$ kg, $m_2 = 172.9 \times 10^3$ kg, $m_3 = 636 \times 10^5$ kg, $d_1 = 215 \times 10^2$ kg·s^{-1}, $d_2 = 97 \times 10^3$ kg·s^{-1} and $d_3 = 802 \times 10^4$ kg·m$^2 \cdot$ s^{-1}. The saturation limits are given as $\tau_{u\,max} = 2 \times 10^6$ and $\tau_{r\,max} = 1.5 \times 10^7$. For the real ship, we consider the following initial position: $(x(0), y(0), \phi(0), u(0), v(0), r(0)) = (8, 7, 0, 0, 0, 0)$. The disturbances are assumed to be $d_u = 100 \times \sin(t)$, $d_v = 50 \times \sin(t)$ and $d_r = 100 \times \cos(t)$. The control parameters are selected as: $c_1 = 4$, $c_2 = 2$, $c_3 = 10$, $c_4 = 4$, $c_5 = 3$, $c_6 = 3$, $c = 2$, $\varepsilon_1 = 0.6$, $\varepsilon_2 = 0.6$, $k_3 = 20001$, $\varepsilon_3 = 0.01$, $\varepsilon_4 = 110$. Simulation time is 10s. Figure 4.7 shows the reference and the trajectory of the real ship. Fig.4.6 shows *rho* and the disturbance $d_1(t)$. It is shown that the disturbance is estimated by observer. The position of the ship without disturbance is shown in Fig.4.7. Figure 4.8 shows the positions of the reference and real ship under disturbances $d_1 = 1000 * \sin(t)$ and $d_2 = 2000 * \cos(t)$. It could be seen that the stabilization control could be achieved and the position errors converge to the origin asymptotically. Figure 4.9 shows the torque τ_u and τ_r. It can be seen that the τ_u and τ_r are strictly smaller than the respective bound.

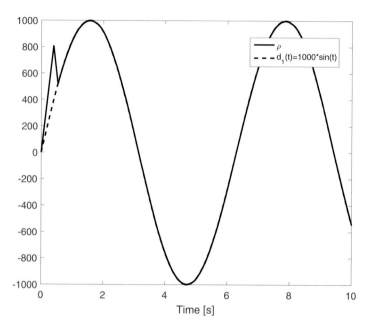

Fig. 4.6. ρ and $d_1(t)$.

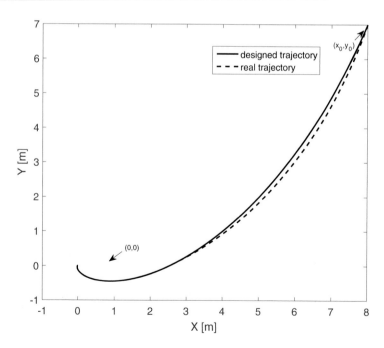

Fig. 4.7. The positions of the reference and real ship without disturbance.

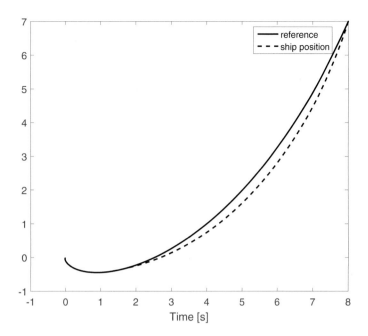

Fig. 4.8. The positions of the reference and real ship under disturbances $d_1 = 1000 * \sin(t)$ and $d_2 = 2000 * \cos(t)$.

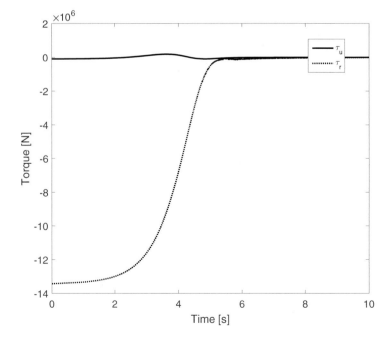

Fig. 4.9. The torque τ_u and τ_r.

4.5 Conclusion

In this chapter, we consider the stabilization control problem of underactuated ships subject to input saturation. First the by introducing a virtual leader, the stabilization of an underactuated ship is transformed into a path following problem thus the error dynamic system can be divided into a cascade of two subsystems, i.e., (ϕ_e, r_e) subsystem and (x_e, y_e, u, v) subsystem. With our proposed finite-time disturbance observer and yaw controller, the two subsystems are de-coupled within finite time. This enables the torque in surge to be designed independently by using a new backstepping technique. It is shown that the system is stable and the position errors converge to the origin asymptotically.

5

Global Adaptive Stabilization Control of Underactuated Ships with Nussbaum Function

In this chapter, global adaptive stabilization control of an underactuated ship in the presence of input saturation and external disturbances is investigated. By proposing a novel Nussbaum-type function, the augmented system approach is extended to the saturated and interconnected nonaffine nonlinear systems and Lyapunov stability is established. The uncertain parameters and external disturbances are not required to be in a known compact set. Globally asymptotically convergence of tracking errors has been established and a simulation example is given to illustrate the effectiveness of the proposed control scheme.

5.1 Introduction

The control of underactuated systems has received lots of attention over the past few decades. Typical examples of such systems include nonholonomic mobile robots, underactuated ships, underwater vehicles and VCTOL aircrafts *et al*. There has been significant interest in control of underactuated ships and various control schemes are proposed for such a problem [42, 20, 16, 19, 22, 18, 21, 47]. Normally the tracking and stabilization control of underactuated ships are investigated separately due to the existence of underactuation, and the control schemes cannot be applied each other. In the tracking control, backstepping and the Lyapunov function approach are used extensively. In [17], two effective control schemes are presented to solve the global tracking control problem for underactuated ships with the Lyapunov method by exploiting the inherent cascade interconnected structure of the ship dynamics. In [18, 21] the tracking or path following control of underactuated ships are solved by backstepping techniques. In [22] a controller that achieves practical tracking of arbitrary reference trajectories for underactuated ships using the transverse function approach. However, the stabilization control of underactuated ships is normally solved by exploring the chained structure of the dynamics of the ship through variable exchange. For example, in [82], the stabilization problem of underactuated ships is reduced to the stabilization of a third-order chained form by a change of coordinates. Subsequent related works on the control of underactuated ships include, but are not limited to, [45, 16, 46, 42, 20] and many references therein.

In practice, an underactuated ship is ultimately driven by a motor which can only provide limited amount of torque. Thus the magnitude of the control signal is always constrained. Under such a constraint, the controllability of the underactuated ships is seriously affected. Thus failing to tackle the input saturation constraint may worsen the control performance, which could be severely limited, and even give rise to undesirable inaccuracy or lead to insta- bility. Therefore, the development of a control scheme for an underactuated ship with input saturation has been a difficult task with practical interest and theoretical significance. The tracking control of underactuated ships with in- put saturation have received attentions and several works have been proposed. For example, in [47] a tracking control method is presented to address such an issue by employing the linearization and dynamic surface control, only en- suring bounded tracking errors. In [83] tracking control of an underactuated ship in the presence of input saturation is addressed by dividing the track- ing error dynamic system into a cascade of two subsystems. The torques in surge and yaw axis are designed separately using the backstepping technique. In [84] a global state feedback tracking controller for underactuated surface marine vessels is proposed under an assumption on the reference trajectory. However, all these results are not satisfactory in several respects; for example, the system parameters are required to be known, the external disturbances are required to be zero and the tracking errors could not be made arbitrarily small, etc. In this chapter, an adaptive stabilization controller is proposed for an underactuated ship in the presence of parametric uncertainties and external disturbances with input saturation, based on the Nussbaum func- tion approach, guaranteeing global stability and that the tracking errors are arbitrarily small.

5.2 A Novel Nassbaum Function and A Key Lemma

A Nussbaum-type function $\mathcal{N}(\cdot)$ has the following properties:

$$\lim_{k \to \infty} \sup \frac{1}{k} \int_0^k \mathcal{N}(\tau) d\tau = +\infty,$$

$$\lim_{k \to \infty} \inf \frac{1}{k} \int_0^k \mathcal{N}(\tau) d\tau = -\infty. \tag{5.1}$$

Commonly used Nussbaum-type functions include $k^2 \cos(k)$, $k^2 \sin(k)$, $e^{k^2} \cos(k)$, et al which is shown in Fig.5.1. However, normally these exist- ing Nussbaum-type functions could not applied to the control problems such as interconnected nonlinear systems and MIMO nonlinear systems directly,

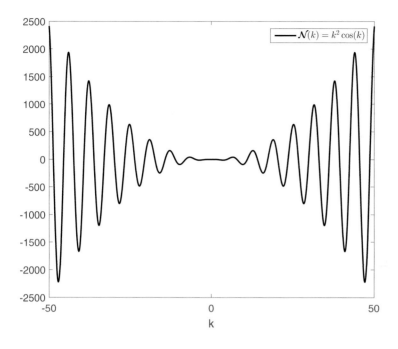

Fig. 5.1. A common Nussbaum function.

especially the interconnected high-order nonlinear systems considered in this chapter. This is because it is still unclear how to solve the coexistence of multiple Nussbaum-type functions such as $k^2 \cos(k)$ in a single Lyapunov inequality, as will be shown later. To overcome this obstacle, we propose a new Nussbaum-type function for each subsystem.

$$\mathcal{N}_i(k_1, ..., k_N) = k_i(k_1^2 + ... + k_N^2)^2 \sin(k_1^2 + ... + k_N^2) \qquad (5.2)$$

where $i = 1, ..., N$. Now we establish a key lemma for $\mathcal{N}_i(k_1, ..., k_N)$ as follows.

Lemma 5.1. *If there exists a positive definite, radially unbounded function $V(t)$ satisfying the following inequality,*

$$V(t) \leq \sum_{i=1}^{N} \int_0^t \mathcal{N}_i(k_1, ..., k_N)\dot{k}_i(\tau)d\tau - \sum_{i=1}^{N} \int_0^t \dot{k}_i(\tau)d\tau + c \qquad (5.3)$$

for $t \in [0, t_f)$ where c is a constant, then $V(t)$, k_i and $\sum_{i=1}^{N} \int_0^t \mathcal{N}_i(k_1, ..., k_N)\dot{k}_i(\tau)d\tau$ are bounded on $[0, t_f)$.

Proof: From (5.3) we obtain

$$
V(t) \leq \sum_{i=1}^{N} \int_0^t \mathcal{N}_i(k_1,...,k_N)\dot{k}_i(\tau)d\tau - \sum_{i=1}^{N} \int_0^t \dot{k}_i(\tau)d\tau + c
$$

$$
= \int_0^t (k_1^2 + ... + k_N^2)^2 \sin\left(k_1^2 + ... + k_N^2\right) \sum_{i=1}^{N} k_i \dot{k}_i(\tau)d\tau
$$

$$
- \sum_{i=1}^{N} k_i(t) + c + \sum_{i=1}^{N} k_i(0)
$$

$$
= \frac{1}{2} \int_{k_1(0)^2 + ... + k_N(0)^2}^{k_1^2 + ... + k_N^2} \sigma^2 \sin(\sigma)d\sigma - \sum_{i=1}^{N} k_i(t) + c + \sum_{i=1}^{N} k_i(0)
$$

$$
= \frac{1}{2}\left(-\lambda^2 \cos(\lambda) + 2\lambda \sin(\lambda) + 2\cos(\lambda)\right) - \sum_{i=1}^{N} k_i(t) + \bar{c} \qquad (5.4)
$$

where $\lambda(t) = k_1(t)^2 + ... + k_N(t)^2$ and $\bar{c} = c + \sum_{i=1}^{N} k_i(0) - \frac{1}{2}(-\lambda(0)^2 \cos(\lambda(0)) + 2\lambda(0)\sin(\lambda(0)) + 2\cos(\lambda(0)))$ is a constant.

Then we will prove the boundedness of $V(t)$, k_i and $\sum_{i=1}^{N} \int_0^t \mathcal{N}_i(k_1,...,k_N)$ $\dot{k}_i(\tau)d\tau$ by seeking a contradiction. Suppose some $k_i(t)$ are unbounded, then $\lambda(t)$ is also unbounded. Let $\lambda(t^*) = 2n\pi$ where $n \in \mathcal{N}^+ \to +\infty$ for $t = t^*$, then from (5.4) it is easy to see that $V(t^*) \to -\infty$, which makes a contradiction. Thus $k_i(t)$, $i = 1, ..., N$ are bounded on $t \in [0, t_f)$ for any $t_f > 0$, which further implies $V(t)$ and $\sum_{i=1}^{N} \int_0^t \mathcal{N}_i(k_1,...,k_N)\dot{k}_i(\tau)d\tau$ are also bounded on $[0, t_f)$.

5.3 Problem Formulation and Controller Design

5.3.1 Problem Formulation

We consider the ship as shown in Fig.5.2 where the dynamic model can be described as follows

$$
\dot{\bar{x}} = \cos(\bar{\phi})u - \sin(\bar{\phi})v
$$

$$
\dot{\bar{y}} = \sin(\bar{\phi})u + \cos(\bar{\phi})v
$$

$$
\dot{\bar{\phi}} = r
$$

$$
\dot{u} = \frac{m_{22}}{m_{11}}vr - \frac{d_{11}}{m_{11}}u - \sum_{i=2}^{3} \frac{d_{ui}}{m_{11}}|u|^{i-1}u + \frac{1}{m_{11}}\mathrm{sat}(\bar{\tau}_u) + \frac{1}{m_{11}}\tau_{wu}(t)
$$

$$
\dot{v} = -\frac{m_{11}}{m_{22}}ur - \frac{d_{22}}{m_{22}}v - \sum_{i=2}^{3} \frac{d_{vi}}{m_{22}}|v|^{i-1}v + \frac{1}{m_{22}}\tau_{wv}(t)
$$

$$
\dot{r} = \frac{m_{11} - m_{22}}{m_{33}}uv - \frac{d_{33}}{m_{33}}r - \sum_{i=2}^{3} \frac{d_{ri}}{m_{33}}|r|^{i-1}r + \frac{1}{m_{33}}\mathrm{sat}(\bar{\tau}_r) + \frac{1}{m_{33}}\tau_{wr}(t)
$$

$$
(5.5)
$$

Fig. 5.2. A real ship model.

where $\bar{x}, \bar{y}$ denote the position of the ship surface in X and Y directions, respectively. $\bar{\phi}$ denotes the heading angle of the ship. u, v and r are the surge, sway and yaw velocities, respectively. τ_u and τ_r represent the surge and yaw forces, respectively, which are considered as actual control inputs where the saturation bounds are τ_{um} and τ_{rm} respectively. Since there is no force introduced in the sway direction, the system modeled by (5.5) is underactuated. The positive constants m_{jj} denote the ship inertia including added mass, and d_{jj}, d_{ui}, d_{vi}, d_{ri} for $2 \leq i \leq 3$, $1 \leq j \leq 3$ represent the hydrodynamic damping coefficients. Note that in this chapter, all these constants are assumed to be unknown. The time-varying terms, $\tau_{wu}(t)$, $\tau_{wv}(t)$ and $\tau_{wr}(t)$, are the environmental disturbances induced by the wave, wind and ocean-current.

Design the control inputs τ_u and τ_r to force the ship, as modeled (5.5), to follow a prescribed path set, which is denoted by $\Omega = (x_d(s), y_d(s))$ with s being a path parameter. To achieve this objective, the following assumption is imposed.

Assumption 5.1 $x_d'(s)^2 + y_d'(s)^2 \geq \mu$ where μ is a positive constant, $x_d'(s) = \dfrac{\partial x_d(s)}{\partial s}$ and $y_d'(s) = \dfrac{\partial y_d(s)}{\partial s}$.

Assumption 5.2 *The minimum radius of the osculating circle of the path is larger than or equal to the minimum possible turning radius of the ships.*

5.3.2 Ship Dynamics Transformation

In order to tackle the underactuated problem, a coordinate change is firstly performed as a preliminary step. The transverse function approach is adopted to introduce an "auxiliary manipulated variable" in addition to the actual control inputs.

- *Change of coordinates:*

$$\begin{bmatrix} x \\ y \end{bmatrix} = \begin{bmatrix} \bar{x} \\ \bar{y} \end{bmatrix} + R(\phi) \begin{bmatrix} f_1(\alpha) \\ f_2(\alpha) \end{bmatrix},$$
$$\phi = \bar{\phi} - f_3(\alpha) \tag{5.6}$$

where $R(\phi) = [\cos(\phi), -\sin(\phi); \sin(\phi), \cos(\phi)]$ and $f_l(\alpha)$ for $l = 1, 2, 3$ are three functions of α which will be determined later. Taking the derivatives of x, y and ϕ gives

$$\begin{bmatrix} \dot{x} \\ \dot{y} \end{bmatrix} = Q \begin{bmatrix} u \\ \dot{\alpha} \end{bmatrix} + \begin{bmatrix} -v \sin(\bar{\phi}) \\ v \cos(\bar{\phi}) \end{bmatrix}$$
$$+ R'(\phi) \begin{bmatrix} f_1(\alpha) \\ f_2(\alpha) \end{bmatrix} (r - f_3'(\alpha)\dot{\alpha}) \tag{5.7}$$

$$\dot{\phi} = r - f_3'\dot{\alpha} \tag{5.8}$$

where $\dot{\alpha}$ is deemed as an auxiliary manipulated variable, and

$$Q = \left[\begin{bmatrix} \cos(\bar{\phi}) \\ \sin(\bar{\phi}) \end{bmatrix}, \quad R(\phi) \begin{bmatrix} f_1'(\alpha) \\ f_2'(\alpha) \end{bmatrix} \right],$$

$f_l' = \dfrac{\partial f_l(\alpha)}{\partial \alpha}$ for $l = 1, 2, 3$ and $R'(\phi) = \dfrac{\partial R(\phi)}{\partial \phi}$. $f_l(\alpha)$ are chosen as follows such that Q is ensured to be invertible for all $\bar{\phi} \in \Re$ and $\alpha \in \Re$.

$$f_1(\alpha) = \epsilon_1 \sin(\alpha)\frac{\sin(f_3)}{f_3}, f_2(\alpha) = \epsilon_1 \sin(\alpha)\frac{1 - \cos(f_3)}{f_3},$$
$$f_3(\alpha) = \epsilon_2 \cos(\alpha) \tag{5.9}$$

where ϵ_1 and ϵ_2 are constants selected to satisfy that

$$0 < \epsilon_1, \quad 0 < \epsilon_2 < \frac{\pi}{2}.$$

It follows that $|f_1| < \epsilon_1, |f_2| < \epsilon_1, |f_3| < \epsilon_2$ and

$$\det(Q) = \frac{\epsilon_1\epsilon_2}{(\epsilon_2 \cos(\alpha))^2}(\cos(\epsilon_2 \cos(\alpha)) - 1)$$
$$\leq -\frac{\epsilon_1}{\epsilon_2}(1 - \cos(\epsilon_2)) < 0. \tag{5.10}$$

5.3.3 Controller Design

The final controls are $\bar{\tau}_u = g(\tau_u)$ and $\bar{\tau}_r = g(\tau_r)$ where τ_u and τ_r are outputs of two filters

$$\dot{\tau}_u = -c_1\tau_u + \omega_u,$$
$$\dot{\tau}_r = -c_2\tau_r + \omega_r. \tag{5.11}$$

Step 1: Define the path tracking errors as

$$x_e = x - x_d, \quad y_e = y - y_d, \quad \phi_e = \phi - \phi_d. \tag{5.12}$$

The reference angle ϕ_d is given as $\phi_d = \arctan(\frac{y'_d}{x'_d})$ where $x'_d = \dfrac{\partial x_d}{\partial s}$ and $y'_d = \dfrac{\partial y_d}{\partial s}$. Choose the Lyapunov function candidate in this step as

$$V_1 = \frac{1}{2} q_e^T q_e + \frac{1}{2} \phi_e^2 \tag{5.13}$$

where $q_e = [x_e, y_e]^T$. Then the derivative of V_1 can be computed as

$$\dot{V}_1 = q_e^T \left(Q \begin{bmatrix} u \\ \dot{\alpha} \end{bmatrix} + \begin{bmatrix} -v\sin(\bar{\phi}) \\ v\cos(\bar{\phi}) \end{bmatrix} + R'(\phi) \begin{bmatrix} f_1(\alpha) \\ f_2(\alpha) \end{bmatrix} \right.$$

$$\left. \times (r - f'_3(\alpha)\dot{\alpha}) - \dot{q}_d \right) + \phi_e(r - f'_3\dot{\alpha} - \dot{\phi}_d).$$

In the above expression, $q_d = [x_d, y_d]^T$ and $\dot{q}_d = [x'_d(s)\dot{s}, y'_d(s)\dot{s}]^T$, $\dot{\phi}_d = \dfrac{x'_d(s)y''_d(s) - x''_d(s)y'_d(s)}{x'_d(s)^2 + y'_d(s)^2}\dot{s}$, where $x''_d(s) = \frac{\partial^2 x_d}{\partial s^2}$, $y''_d(s) = \frac{\partial^2 y_d}{\partial s^2}$. $\dot{s}$ is selected to be

$$\dot{s} = \frac{u^\star(1 - \epsilon_3 e^{-\epsilon_4 t})e^{-\epsilon_5\|q - q_d\|}}{\sqrt{x'^2_d + y'^2_d}} \tag{5.14}$$

where $u^\star$ is a non-zero prescribed speed, $\epsilon_i > 0$ for $i = 3, 4, 5$, $\epsilon_3 < 1$ and $q = [x, y]^T$.

Introduce two new error variables

$$u_e = u - u_d, r_e = r - r_d \tag{5.15}$$

where u_d and r_d are the virtual controls for u and r, respectively. u_d, r_d and $\dot{\alpha}$ are chosen as

$$\begin{bmatrix} u_d \\ \dot{\alpha} \end{bmatrix} = Q^{-1} \left(-k_1 q_e - \begin{bmatrix} -v\sin(\bar{\phi}) \\ v\cos(\bar{\phi}) \end{bmatrix} - R'(\phi) \begin{bmatrix} f_1(\alpha) \\ f_2(\alpha) \end{bmatrix} \right.$$

$$\left. \times (-k_2\phi_e + \dot{\phi}_d) + \dot{q}_d \right) \tag{5.16}$$

$$r_d = -k_2\phi_e + f'_3\dot{\alpha} + \dot{\phi}_d \tag{5.17}$$

where k_1 and k_2 are positive constants.

From (5.16) and (5.17), we obtain that

$$\dot{V}_1 = -k_1 q_e^T q_e - k_2 \phi_e^2 + \varrho_u u_e + \varrho_r r_e$$

where $\varrho_u = q_e^T Q[1,\ 0]^T$ and $\varrho_r = q_e^T R'(\phi)[f_1(\alpha),\ f_2(\alpha)]^T + \phi_e$.

Step 2: Let $z_1 = g(\tau_u) - \alpha_u$ and $z_2 = g(\tau_r) - \alpha_r$. Taking derivatives of both sides of (5.15) yields

$$
\dot{u}_e = \frac{m_{22}}{m_{11}} vr - \frac{d_{11}}{m_{11}} u - \sum_{i=2}^{3} \frac{d_{ui}}{m_{11}} |u|^{i-1} u + \frac{1}{m_{11}} (z_1 + \alpha_u)
$$
$$
+ \frac{1}{m_{11}} \tau_{wu}(t) - \frac{\partial u_d}{\partial x_e} \dot{x}_e - \frac{\partial u_d}{\partial y_e} \dot{y}_e - \frac{\partial u_d}{\partial s} \dot{s} - \frac{\partial u_d}{\partial \phi} \dot{\phi}
$$
$$
- \frac{\partial u_d}{\partial \bar{\phi}} \dot{\bar{\phi}} - \frac{\partial u_d}{\partial \alpha} \dot{\alpha} - \frac{\partial u_d}{\partial v} \left(-\frac{m_{11}}{m_{22}} ur - \frac{d_{22}}{m_{22}} v \right.
$$
$$
\left. - \sum_{i=2}^{3} \frac{d_{vi}}{m_{22}} |v|^{i-1} v + \frac{1}{m_{22}} \tau_{wv}(t) \right) - \frac{\partial u_d}{\partial x'_d} \ddot{x}_d - \frac{\partial u_d}{\partial y'_d} \ddot{y}_d - \frac{\partial u_d}{\partial \phi'_d} \ddot{\phi}_d,
$$

$$
\dot{r}_e = \frac{m_{11} - m_{22}}{m_{33}} uv - \frac{d_{33}}{m_{33}} r - \sum_{i=2}^{3} \frac{d_{ri}}{m_{33}} |r|^{i-1} r + \frac{1}{m_{33}} (z_2 + \alpha_r)
$$
$$
+ \frac{1}{m_{33}} \tau_{wr}(t) - \frac{\partial r_d}{\partial x_e} \dot{x}_e - \frac{\partial r_d}{\partial y_e} \dot{y}_e - \frac{\partial r_d}{\partial s} \dot{s} - \frac{\partial r_d}{\partial \phi} \dot{\phi}
$$
$$
- \frac{\partial r_d}{\partial \bar{\phi}} \dot{\bar{\phi}} - \frac{\partial r_d}{\partial \alpha} \dot{\alpha} - \frac{\partial r_d}{\partial v} \left(-\frac{m_{11}}{m_{22}} ur - \frac{d_{22}}{m_{22}} v - \sum_{i=2}^{3} \frac{d_{vi}}{m_{22}} |v|^{i-1} v \right.
$$
$$
\left. + \frac{1}{m_{22}} \tau_{wv}(t) \right) - \frac{\partial r_d}{\partial x'_d} \ddot{x}_d - \frac{\partial r_d}{\partial y'_d} \ddot{y}_d - \frac{\partial r_d}{\partial \phi'_d} \ddot{\phi}_d.
$$

We choose the Lyapunov function candidate in this step as

$$
V_2 = V_1 + \frac{m_{11}}{2} u_e^2 + \frac{m_{33}}{2} r_e^2 + \frac{1}{2} \sum_{i=1}^{3} \tilde{\theta}_i^T \Gamma_i^{-1} \tilde{\theta}_i \tag{5.18}
$$

where $\Gamma_i = \mathrm{diag}(\delta_{ij})$ for $i = 1, 2, 3$ are positive definite matrices with appropriate dimensions. $\tilde{\theta}_i = \theta_i - \hat{\theta}_i$ denote the estimation errors. $\hat{\theta}_i$ is the estimate designed for θ_i, where

$$
\theta_1 = \left[m_{22}, d_u, d_{u2}, d_{u3}, m_{11}, \frac{m_{11}^2}{m_{22}}, \frac{d_v m_{11}}{m_{22}}, \frac{d_{v2} m_{11}}{m_{22}}, \frac{d_{v3} m_{11}}{m_{22}} \right],
$$
$$
\theta_2 = \left[(m_{11} - m_{22}), d_r, d_{r2}, d_{r3}, m_{33}, \frac{m_{11} m_{33}}{m_{22}}, \frac{d_v m_{33}}{m_{22}}, \frac{d_{v2} m_{33}}{m_{22}}, \frac{d_{v3} m_{33}}{m_{22}} \right],
$$
$$
\theta_3 = \left[\tau_{wu\,\max}, \frac{m_{11}}{m_{22}} \tau_{wv\,\max}, \tau_{wr\,\max}, \frac{m_{33}}{m_{22}} \tau_{wv\,\max} \right].
$$

Without canceling the useful damping term, the actual controls for τ_u and τ_r are designed as

$$\alpha_u = -k_3 u_e - \hat{\theta}_1^T \beta_1 + \varrho_u - \hat{\theta}_{31} \tanh(\frac{u_e}{\varepsilon_1}) - \hat{\theta}_{32} \frac{\partial u_d}{\partial v} \tanh(\frac{\partial u_d}{\partial v} \frac{u_e}{\varepsilon_2}),$$

$$\alpha_r = -k_4 r_e - \hat{\theta}_2^T \beta_2 + \varrho_r - \hat{\theta}_{33} \tanh(\frac{r_e}{\varepsilon_3}) - \hat{\theta}_{34} \frac{\partial r_d}{\partial v} \tanh(\frac{\partial r_d}{\partial v} \frac{r_e}{\varepsilon_4}) \quad (5.19)$$

where k_3, k_4, ε_i, $1 \le i \le 4$ are positive constants and

$$\beta_1 = \left[vr, -u_d, -|u|u_d, -u^2 u_d, \gamma_1, \frac{\partial u_d}{\partial v} ur, \frac{\partial u_d}{\partial v} v, \right.$$
$$\left. \frac{\partial u_d}{\partial v} |v|v, \frac{\partial u_d}{\partial v} v^3 \right],$$

$$\beta_2 = \left[uv, -r_d, -|r|r_d, -r^2 r_d, \gamma_2, \frac{\partial r_d}{\partial v} ur, \frac{\partial r_d}{\partial v} v, \right.$$
$$\left. \frac{\partial r_d}{\partial v} |v|v, \frac{\partial r_d}{\partial v} v^3 \right],$$

$$\gamma_1 = -\frac{\partial u_d}{\partial x_e} \dot{x}_e - \frac{\partial u_d}{\partial y_e} \dot{y}_e - \frac{\partial u_d}{\partial s} \dot{s} - \frac{\partial u_d}{\partial \phi} \dot{\phi} - \frac{\partial u_d}{\partial \bar{\phi}} \dot{\bar{\phi}}$$
$$- \frac{\partial u_d}{\partial \alpha} \dot{\alpha} - \frac{\partial u_d}{\partial x_d'} \ddot{x}_d - \frac{\partial u_d}{\partial y_d'} \ddot{y}_d - \frac{\partial u_d}{\partial \phi_d'} \ddot{\phi}_d,$$

$$\gamma_2 = -\frac{\partial r_d}{\partial x_e} \dot{x}_e - \frac{\partial r_d}{\partial y_e} \dot{y}_e - \frac{\partial r_d}{\partial s} \dot{s} - \frac{\partial r_d}{\partial \phi} \dot{\phi} - \frac{\partial r_d}{\partial \bar{\phi}} \dot{\bar{\phi}}$$
$$- \frac{\partial r_d}{\partial \alpha} \dot{\alpha} - \frac{\partial r_d}{\partial x_d'} \ddot{x}_d - \frac{\partial r_d}{\partial y_d'} \ddot{y}_d - \frac{\partial r_d}{\partial \phi_d'} \ddot{\phi}_d.$$

The adaptive laws are chosen as $\dot{\hat{\theta}}_1 = u_e \Gamma_1 \beta_1$, $\dot{\hat{\theta}}_2 = r_e \Gamma_2 \beta_2$, $\dot{\hat{\theta}}_{31} = \delta_{31} \tanh(\frac{u_e}{\varepsilon_1})$, $\dot{\hat{\theta}}_{32} = \delta_{32} u_e \frac{\partial u_d}{\partial v} \tanh(\frac{\partial u_d}{\partial v} \frac{u_e}{\varepsilon_2})$, $\dot{\hat{\theta}}_{33} = \delta_{33} \tanh(\frac{r_e}{\varepsilon_3})$, $\dot{\hat{\theta}}_{34} = \delta_{34} r_e \frac{\partial r_d}{\partial v} \tanh(\frac{\partial r_d}{\partial v} \frac{r_e}{\varepsilon_4})$.
Then the time-derivative of the V_2 is

$$\dot{V}_2(t) \le -k_1 q_e^T q_e - k_2 \phi_e^2 - (k_3 + d_u) u_e^2 - (k_4 + d_r) r_e^2$$
$$+ \sum_{i=1}^{4} \Theta_{3i} \delta \varepsilon_i + u_e z_1 + r_e z_2 \quad (5.20)$$

where Θ_{3i} is the bound of θ_{3i}.

Step 3: Taking the time-derivatives of z_1 and z_2 yield

$$\dot{z}_1 = \chi_u(-c_1\tau_u + \omega_u) + \bar{\gamma}_1 - \frac{\partial\alpha_u}{\partial\hat{\theta}_1}\dot{\hat{\theta}}_1 - \sum_{i=1}^{2}\frac{\partial\alpha_u}{\partial\hat{\theta}_{3i}}\dot{\hat{\theta}}_{3i} - \beta_3\theta_4$$

$$- \frac{\partial\alpha_u}{\partial u}\frac{1}{m_{11}}\tau_{wu}(t) - \frac{\partial\alpha_u}{\partial v}\frac{1}{m_{22}}\tau_{wv}(t) - \frac{\partial\alpha_u}{\partial r}\frac{1}{m_{33}}\tau_{wr}(t),$$

$$\dot{z}_2 = \chi_r(-c_2\tau_r + \omega_r) + \bar{\gamma}_2 - \frac{\partial\alpha_r}{\partial\hat{\theta}_2}\dot{\hat{\theta}}_2 - \sum_{i=3}^{4}\frac{\partial\alpha_r}{\partial\hat{\theta}_{3i}}\dot{\hat{\theta}}_{3i} - \beta_4\theta_4$$

$$- \frac{\partial\alpha_r}{\partial u}\frac{1}{m_{11}}\tau_{wu}(t) - \frac{\partial\alpha_r}{\partial v}\frac{1}{m_{22}}\tau_{wv}(t) - \frac{\partial\alpha_r}{\partial r}\frac{1}{m_{33}}\tau_{wr}(t) \qquad (5.21)$$

where $\bar{\gamma}_1$ and $\bar{\gamma}_2$ are defined similar to γ_1 and γ_2 except that u_d and r_d are replaced by α_u and α_r respectively, and

$$\beta_3 = \left[\frac{\partial\alpha_u}{\partial u}vr, -\frac{\partial\alpha_u}{\partial u}u, -\frac{\partial\alpha_u}{\partial u}|u|u, -\frac{\partial\alpha_u}{\partial u}u^3, \frac{\partial\alpha_u}{\partial u}g(\tau_u), \frac{\partial\alpha_u}{\partial v}ur, \frac{\partial\alpha_u}{\partial v}v, \right.$$
$$\left. \frac{\partial\alpha_u}{\partial v}|v|v, \frac{\partial\alpha_u}{\partial v}v^3, \frac{\partial\alpha_u}{\partial r}uv, -\frac{\partial\alpha_u}{\partial r}r, -\frac{\partial\alpha_u}{\partial r}|r|r, -\frac{\partial\alpha_u}{\partial r}r^3, \frac{\partial\alpha_u}{\partial r}g(\tau_r)\right],$$

$$\beta_4 = \left[\frac{\partial\alpha_r}{\partial u}vr, -\frac{\partial\alpha_r}{\partial u}u, -\frac{\partial\alpha_r}{\partial u}|u|u, -\frac{\partial\alpha_r}{\partial u}u^3, \frac{\partial\alpha_r}{\partial u}g(\tau_u), \frac{\partial\alpha_r}{\partial v}ur, \frac{\partial\alpha_u}{\partial v}v, \right.$$
$$\left. \frac{\partial\alpha_r}{\partial v}|v|v, \frac{\partial\alpha_r}{\partial v}v^3, \frac{\partial\alpha_r}{\partial r}uv, -\frac{\partial\alpha_r}{\partial r}r, -\frac{\partial\alpha_r}{\partial r}|r|r, -\frac{\partial\alpha_r}{\partial r}r^3, \frac{\partial\alpha_r}{\partial r}g(\tau_r)\right],$$

$$\theta_4 = \left[\frac{m_{22}}{m_{11}}, \frac{d_u}{m_{11}}, \frac{d_{u2}}{m_{11}}, \frac{d_{u3}}{m_{11}}, \frac{1}{m_{11}}, -\frac{m_{11}}{m_{22}}, -\frac{d_v}{m_{22}}, \right.$$
$$\left. -\frac{d_{v2}}{m_{22}}, -\frac{d_{v3}}{m_{22}}, \frac{m_{11}-m_{22}}{m_{33}}, -\frac{d_r}{m_{33}}, -\frac{d_{r2}}{m_{33}}, -\frac{d_{r3}}{m_{33}}, -\frac{1}{m_{33}}\right],$$

$$\theta_5 = \left[\frac{1}{m_{11}}\tau_{wu\,max}, \frac{1}{m_{22}}\tau_{wv\,max}, \frac{1}{m_{33}}\tau_{wr\,max}\right].$$

The virtual control ω_u and ω_r are designed as

$$\omega_u = \frac{1}{\chi_u}\mathcal{N}_u(k_u, k_r)\bar{\omega}_u,$$

$$\bar{\omega}_u = -k_5z_1 + \chi_uc_1\tau_u + u_e + \beta_3\hat{\theta}_{4u} - \bar{\gamma}_1 + \frac{\partial\alpha_u}{\partial\hat{\theta}_1}\dot{\hat{\theta}}_1 + \sum_{i=1}^{2}\frac{\partial\alpha_u}{\partial\hat{\theta}_{3i}}\dot{\hat{\theta}}_{3i}$$

$$- \hat{\theta}_{51u}\frac{\partial\alpha_u}{\partial u}\tanh(\frac{\partial\alpha_u}{\partial u}\frac{z_1}{\varepsilon_1}) - \hat{\theta}_{52u}\frac{\partial\alpha_u}{\partial v}\tanh(\frac{\partial\alpha_u}{\partial v}\frac{z_1}{\varepsilon_2}) - \hat{\theta}_{53u}\frac{\partial\alpha_u}{\partial r}\tanh(\frac{\partial\alpha_u}{\partial r}\frac{z_1}{\varepsilon_3}),$$

$$\omega_r = \frac{1}{\chi_r}\mathcal{N}_r(k_u, k_r)\bar{\omega}_r,$$

$$\bar{\omega}_r = -k_6z_2 + \chi_rc_2\tau_r + r_e + \beta_4\hat{\theta}_{4r} - \bar{\gamma}_2 + \frac{\partial\alpha_r}{\partial\hat{\theta}_2}\dot{\hat{\theta}}_2 + \sum_{i=1}^{2}\frac{\partial\alpha_r}{\partial\hat{\theta}_{3i}}\dot{\hat{\theta}}_{3i}$$

$$- \hat{\theta}_{51r}\frac{\partial\alpha_r}{\partial u}\tanh(\frac{\partial\alpha_u}{\partial u}\frac{z_2}{\varepsilon_1}) - \hat{\theta}_{52r}\frac{\partial\alpha_r}{\partial v}\tanh(\frac{\partial\alpha_u}{\partial v}\frac{z_2}{\varepsilon_2}) - \hat{\theta}_{53r}\frac{\partial\alpha_r}{\partial r}\tanh(\frac{\partial\alpha_u}{\partial r}\frac{z_2}{\varepsilon_3})$$

$$\qquad (5.22)$$

and the parameter estimates are designed as:

$$\dot{\theta}_{4u} = z_1 \Gamma_5 \beta_3, \quad \dot{\theta}_{4r} = z_2 \Gamma_6 \beta_4,$$

$$\dot{\theta}_{51u} = \delta_{71} z_1 \frac{\partial \alpha_u}{\partial u} \tanh(\frac{\partial \alpha_u}{\partial u} \frac{z_1}{\varepsilon_1}), \quad \dot{\theta}_{52u} = \delta_{72} z_1 \frac{\partial \alpha_u}{\partial v} \tanh(\frac{\partial \alpha_u}{\partial v} \frac{z_1}{\varepsilon_2}),$$

$$\dot{\theta}_{53u} = \delta_{73} z_1 \frac{\partial \alpha_u}{\partial r} \tanh(\frac{\partial \alpha_u}{\partial r} \frac{z_1}{\varepsilon_3}), \quad \dot{\theta}_{51r} = \delta_{71} z_2 \frac{\partial \alpha_r}{\partial u} \tanh(\frac{\partial \alpha_r}{\partial u} \frac{z_2}{\varepsilon_1}),$$

$$\dot{\theta}_{52r} = \delta_{72} z_2 \frac{\partial \alpha_r}{\partial v} \tanh(\frac{\partial \alpha_r}{\partial v} \frac{z_2}{\varepsilon_2}), \quad \dot{\theta}_{53r} = \delta_{73} z_2 \frac{\partial \alpha_r}{\partial r} \tanh(\frac{\partial \alpha_r}{\partial r} \frac{z_2}{\varepsilon_3}). \tag{5.23}$$

Choose a Lyapunov function

$$V_3 = V_2 + \frac{1}{2} z_1^2 + \frac{1}{2} z_2^2 + \frac{1}{2} \tilde{\theta}_{4u}^T \Gamma_5^{-1} \tilde{\theta}_{4u}$$

$$+ \frac{1}{2} \tilde{\theta}_{4r}^T \Gamma_6^{-1} \tilde{\theta}_{4r} + \frac{1}{2} \tilde{\theta}_{5u}^T \Gamma_7^{-1} \tilde{\theta}_{5u} + \frac{1}{2} \tilde{\theta}_{5r}^T \Gamma_7^{-1} \tilde{\theta}_{5r} \tag{5.24}$$

and let

$$\dot{k}_u = z_1 \bar{\omega}_u, \quad \dot{k}_r = z_2 \bar{\omega}_r \tag{5.25}$$

then

$$\dot{V}_3 \leq \left[(\mathcal{N}_u(k_u, k_r) - 1)\dot{k}_u + (\mathcal{N}_r(k_u, k_r) - 1)\dot{k}_r \right] - k_1 q_e^T q_e - k_2 \phi_e^2$$

$$- (k_3 + d_u) u_e^2 - (k_4 + d_r) r_e^2 - k_5 z_1^2 - k_6 z_2^2 + c\varepsilon \tag{5.26}$$

where $c = \max\{\delta \Theta_{31}, ..., \delta \Theta_{34}, 2\delta \Theta_{51}, ..., 2\delta \Theta_{53}\}$. Based on Lemma 5.1, the following theorem is given.

Theorem 5.2. *Consider the underactuated ship system (5.5), with the controller (5.22) and parameter update laws (5.23) under Assumption 5.1 and Assumption 5.2. All the signals in the closed loop system are bounded, and the reference trajectories $x_d(s)$ and $y_d(s)$ can be tracked with bounded tracking errors.*

Proof: Since the ship is with input saturation, then based on Lemma 5.1 of [83], it is easy to show the boundedness of u, v and r. Due to the projection operation, $\tilde{\theta}_i$, $i = 1, 2, 3$ are bounded. Thus from (5.26), all signals containing in V_3 are bounded. Hence, x_e, y_e and ϕ_e are bounded. From (5.16) and (5.17), it is easy to check that u_d, r_d and $\dot{\alpha}$ are bounded. From (5.22), the boundedness of τ_u and τ_r is concluded.

From (5.6) and (5.9) it is easy to show that

$$\|(x - \bar{x}, y - \bar{y})\| \leq \sqrt{2\epsilon_1^2}, \quad |\phi - \bar{\phi}| \leq \epsilon_2. \tag{5.27}$$

Finally from (5.6), the tracking errors $\bar{x} - x_d$ and $\bar{y} - y_d$ can be shown to satisfy

$$|\bar{x} - x_d| \leq |\bar{x} - x| + |x_e|, \quad |\bar{y} - y_d| \leq |\bar{y} - y| + |y_e|. \tag{5.28}$$

Thus the tracking errors are bounded. By adjusting the parameters ϵ_1, ϵ_2 and ε, the tracking errors could be made arbitrarily small.

5.4 Simulations

In this section, we carry out simulations to demonstrate the effectiveness of our controllers and to validate our constructive methodology for underactuated ships. The parameters of the underactuated ship from [47] are used, namely $m_1 = 120 \times 10^3$ kg, $m_2 = 172.9 \times 10^3$ kg, $m_3 = 636 \times 10^5$ kg, $d_1 = 215 \times 10^2$ kg·s^{-1}, $d_2 = 97 \times 10^3$ kg·s^{-1} and $d_3 = 802 \times 10^4$ kg·m^2 · s^{-1}. The saturation limits are given as $\tau_{um} = 10^3 kN$ and $\tau_{rm} = 10^4 kN$. For the real ship, we consider the following initial position: $(x(0), y(0), \phi(0), u(0), v(0), r(0)) = (60, -20, \frac{\pi}{2}, 0, 0, 0)$. The disturbances are assumed to be $d_u = 10^3 \times \sin(t)$, $d_v = 50 \times 10^2 \sin(2 * t)$ and $d_r = 100 \times 10^3 \cos(t)$. The control parameters are selected as: $k_1 = 4$, $k_2 = 2$, $k_3 = 10$, $k_4 = 4$, $k_5 = 3$, $k_6 = 3$, $\epsilon_1 = 0.1$, $\epsilon_2 = 0.1$, $\varepsilon_i = e^{-2t}$, $i = 1, ..., 4$. Simulation time is 20s. Fig.5.3 shows the position of the ship. Figure 5.4 shows the system output of the ship. Figure 5.5 and Figure 5.6 show τ_u and τ_r. Figure 5.7 shows the parameter estimators $\hat{\theta}_1$. With our proposed scheme, while guaranteeing the control performance, the torques are within the saturation bounds, which is given by the designer. Fig.5.8 shows the torques of the ship without considering the input saturation.

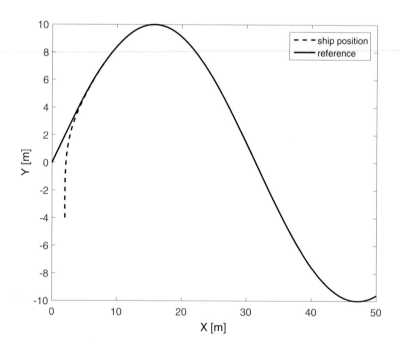

Fig. 5.3. Position of the ship.

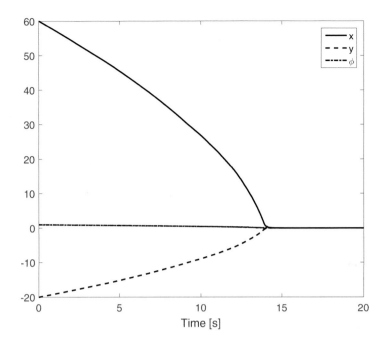

Fig. 5.4. Position of the ship with reference.

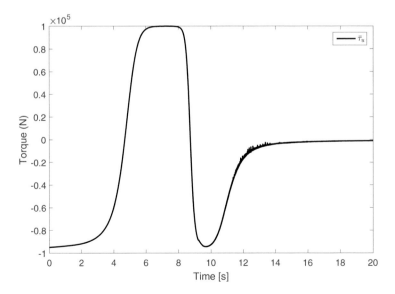

Fig. 5.5. Torque τ_u.

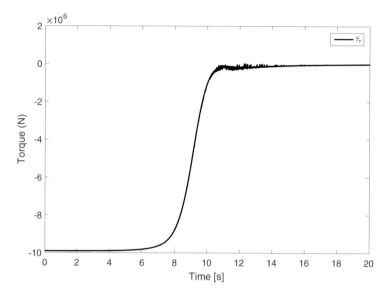

Fig. 5.6. Torque τ_r.

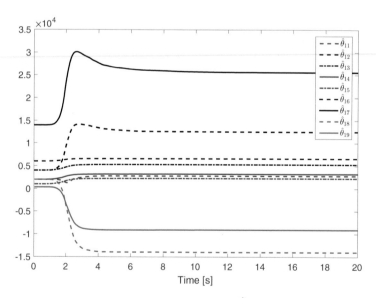

Fig. 5.7. Parameters $\hat{\theta}_1$.

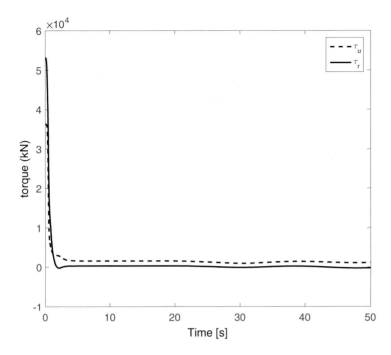

Fig. 5.8. The torque of the ship without considering the input saturation.

5.5 Conclusion

In this chapter, an adaptive control scheme for underactuated ships with input saturation has been proposed. By proposing a novel Nussbaum-type function, the augmented system approach is extended to the saturated underactuated ship system. Globally asymptotically convergence has been proved and a simulation example is given to illustrate the effectiveness of the proposed control scheme.

Part II

Adaptive Output Feedback Control of Underactuated Nonlinear Systems

6

Adaptive Output Feedback Control of Nonholonomic Mobile Robots

In the previous chapters, the tracking or stabilization control of nonholonomic mobile robots and underactuated ships with input saturation are investigated. In the following chapters, adaptive output feedback control of nonholonomic mobile robots and underactuated ships will be addressed.

In this chapter, an adaptive output feedback tracking controller for nonholonomic mobile robots is proposed. The major difficulties of adaptive output feedback for nonholonomic mobile robots are caused by the simultaneous existence of nonholonomic constraints, unknown system parameters and a quadratic term of unmeasurable states in the mobile robot dynamic system as well as their couplings. To overcome these difficulties, a new adaptive control scheme is proposed including designing a new adaptive state feedback controller and two high-gain observers to estimate the unknown linear and angular velocities respectively. It is shown that the closed loop adaptive system is stable and the tracking errors are guaranteed to be within the arbitrarily small pre-specified bounds.

6.1 Introduction

A two-wheeled mobile robot is one of the well-known benchmark nonholonomic systems, which have led to the development of various novel controller design schemes. Much effort has been devoted to the stabilization and tracking control of this kind of nonholonomic systems; for example see [7, 9, 10, 49], and the references therein. In control theory, stabilization is usually regarded as a special case of the tracking problem. However for controlling underactuated mechanical systems with nonholonomic constraints such as nonholonomic mobile robots, the stabilization and tracking problems are totally different, and thus they are normally considered separately. For the stabilization problem, the past abundant literature aims at developing a suitable discontinuous time-invariant stabilizer [3] or time-varying stabilizer [2] or hybrid stabilizers [4, 5]. The seminal work of [2] introduced the first time-varying feedback stabilizer

for a wheeled mobile robot. For tracking control, [6] proposed a time-varying
state feedback tracking controller with backstepping technique [8] for a kinematic model and a simplified dynamic model. [7] designed an adaptive controller based on both kinematic and dynamic models of nonholonomic mobile
robots with unknown parameters. [9] solved both adaptive tracking and stabilization simultaneously. Subsequent related works on the stabilization and
tracking control of nonholonomic mobile robots include, but are not limited
to, [50, 36, 51, 52, 33, 53] and many references therein.

It should be pointed out that most of these proposed controllers are concerned with the case that the velocities of mobile robots are measured by some
devices like tachometers. But in practice such devices may not be used either
because they are contaminated by noise or they are expensive. Many existing
output feedback control schemes such as those proposed in [54, 55] for robot
manipulators cannot be applied to mobile robots due to the nonholonomic
constraint and certain quadratic cross terms. [56] proposed a global output
feedback control for a class of nonholonomic dynamic system, but there is
not any quadratic term of unknown states contained in the system. In [10], a
time-varying output feedback controller was proposed for the dynamic model
of a mobile robot, but it was assumed that all parameters must be known.
In [54], an output feedback controller for single-DOF Lagrangian systems was
proposed with a high-gain observer, but it was difficult to be extended to
systems with more DOFs. To the best of our knowledge, adaptive output
feedback control for nonholonomic mobile robots with unknown parameters
still remains an open research problem. The main difficulty encountered lies in
the simultaneous appearance of the nonholonomic constraint, unknown system parameters and the coriolis matrix, which results in quadratic cross terms
involving unmeasured states. With this in mind, the observer design and the
parameter estimator design are intertwinted, which makes the analysis of the
resulting closed loop system difficult and challenging.

In this chapter, we propose a high-gain observer based scheme to address
the output feedback tracking control of mobile robots in the presence of parametric uncertainties. To accomplish this, we first address the issue of designing
an adaptive state feedback controller with suitable parameter estimators to
achieve asymptotic tracking. Then we design two high-gain observers to estimate the unknown states and substitute the estimates to the state feedback
controller and parameter estimators based on the certainty equivalence prin-
ciple. The stability of the system is analyzed and established by using the
Lyapunov approach. Due to the disturbance rejection property of the high-
gain observers, it is shown that the state estimation errors are of order $O(\varepsilon)$,
where ε is a small design parameter, and the tracking errors can be made
arbitrarily small. With our proposed control scheme, a solution is now presented for the outstanding problem of adaptive output-feedback control for
nonholonomic mobile robots. Finally we demonstrate the effectiveness of our
proposed controllers with simulation studies.

Fig. 6.1. A nonholonomic mobile robot model.

6.2 Robot Model and Problem Formulation

We consider a two-wheeled mobile robot described by the following dynamic models shown in Fig.6.1

$$\dot{\eta} = J(\eta)\omega, \tag{6.1}$$

$$M\dot{\omega} + C(\dot{\eta})\omega + D\omega = \tau, \tag{6.2}$$

where $\eta = (x, y, \phi)$ denotes the position and orientation of the robot, $\omega = (\omega_1, \omega_2)^T$ denotes the angular velocities of the left and right wheels, $\tau = (\tau_1, \tau_2)^T$ represents the control torques applied to the wheels, and M is a symmetric, positive definite inertia matrix, $C(\dot{\eta})$ is the centripetal and coriolis matrix, D denotes the surface friction. Matrices $J(\eta)$, M, $C(\dot{\eta})$ and D are the same as those in Chapter 2. In this chapter, for the sake of simplicity, we assume that the robot does not slip. Also we assume that there is no sliding between the tire and the road, i.e. there is no Coulomb-like friction.

Let v and w denote the linear and angular velocities of the robot, respectively. Then the relationship between ω_1, ω_2 and v, w is described as follows

$$(v, w)^T = B^{-1}(\omega_1, \omega_2)^T, B = \frac{1}{r}\begin{bmatrix} 1 & b \\ 1 & -b \end{bmatrix}. \tag{6.3}$$

Substituting (6.3) into (6.1), we get

$$\frac{d}{dt}\begin{bmatrix} x \\ y \\ \phi \end{bmatrix} = \begin{bmatrix} \cos\phi & 0 \\ \sin\phi & 0 \\ 0 & 1 \end{bmatrix}\begin{bmatrix} v \\ w \end{bmatrix}.$$

For the output feedback control of a nonholonomic mobile robot, η is taken as the output and $\varpi = [v, w]^T$ is the state which is not directly measured in our case. Now let $(x_r, y_r, \phi_r)^T$ denote the desired reference position and orientation of the virtual robot the motion of which is described by:

$$\dot{x}_r = v_r \cos \phi_r,$$
$$\dot{y}_r = v_r \sin \phi_r,$$
$$\dot{\phi}_r = w_r, \tag{6.4}$$

where v_r and w_r denote the linear and angular velocities of the the virtual robot. $(x_r(0), y_r(0))$ denote the initial position and $\phi_r(0)$ is the initial orientation of the reference trajectory. The control objective is to find torque τ applied to the mobile robot to ensure the desired reference trajectories are tracked.

To achieve the objective, we need the following assumptions:

Assumption 6.1 $|x_r(t)| < a^\star, |y_r(t)| < a^\star$, where $a^\star$ is a known but arbitrary positive constant.

Assumption 6.2 The reference linear and angular velocities v_r and w_r and their first-order derivatives are bounded. Also there exist two positive constants $r_\star$ and T_s, such that for $T_s \leq t < \infty$, $r_\star \leq |v_r(t)| < \infty$.

Assumption 6.3 The unknown parameters of the mobile robot are in known compact sets.

6.3 Adaptive State Feedback Control: An Intermediate Step

In this section, a new scheme is proposed to design an adaptive state feedback controller involving a control law and parameter estimators. However, the designed controller will not be implemented as the states are unavailable. Nevertheless, the proposed scheme is helpful for the design of output feedback controller and thus is considered as an "intermediate" step. Basically, the unknown states v and w in the control law and parameter estimators will be replaced by the estimated states $\hat{v}$ and $\hat{w}$ after appropriate modifications in Section IV.

6.3.1 Design of Virtual Control based on Kinematic Model

As we will design high-gain observers to estimate unknown v and w, so unlike the existing schemes such as those in [7, 9], we use v and w as the virtual control variables in state feedback control design. Substituting (6.3) into (6.2), we get

$$\dot{\eta} = L(\eta)\varpi,$$
$$\dot{\varpi} = -B^{-1}M^{-1}C(\dot{\eta})B\varpi - B^{-1}M^{-1}DB\varpi + B^{-1}M^{-1}\tau, \tag{6.5}$$

where

$$L(\eta) = \begin{bmatrix} \cos\phi & 0 \\ \sin\phi & 0 \\ 0 & 1 \end{bmatrix}. \tag{6.6}$$

Let

$$\bar{B} = \frac{1}{r} \begin{bmatrix} -b & -1 \\ -b & 1 \end{bmatrix}.$$

It is obvious that $C(\dot{\eta})B = \bar{B}C(\dot{\eta})$ and thus we can replace $C(\dot{\eta})B$ in (6.5) by $\bar{B}C(\dot{\eta})$. Multiplying (6.5) by $\bar{B}^{-1}MB$, we get

$$\begin{aligned} \dot{\eta} &= L(\eta)\varpi, \\ R\dot{\varpi} &= -\bar{C}(w)\varpi - D_R\varpi + B_R\tau, \end{aligned} \tag{6.7}$$

where $R = -\bar{B}^{-1}MB$, $\bar{C}(w) = -C(w)$, $D_R = -RB^{-1}M^{-1}DB$, $B_R = -RB^{-1}M^{-1}$ and R is a positive definite matrix. Explicitly, we have

$$R = \begin{bmatrix} (m_{11}+m_{22})/b & 0 \\ 0 & (m_{11}-m_{22})b \end{bmatrix}, \quad B_R = \frac{r}{2b}\begin{bmatrix} 1 & 1 \\ b & -b \end{bmatrix},$$

$$D_R = \begin{bmatrix} \frac{d_{11}+d_{22}}{2b} & \frac{d_{11}-d_{22}}{2} \\ \frac{d_{11}-d_{22}}{2} & \frac{b(d_{11}+d_{22})}{2} \end{bmatrix}. \tag{6.8}$$

As often done in tracking control of mobile robots, see for examples [6, 7, 9], the tracking errors are defined as

$$\begin{bmatrix} x_e \\ y_e \\ \phi_e \end{bmatrix} = \begin{bmatrix} \cos\phi & \sin\phi & 0 \\ -\sin\phi & \cos\phi & 0 \\ 0 & 0 & 1 \end{bmatrix} \begin{bmatrix} x - x_r \\ y - y_r \\ \phi - \phi_r \end{bmatrix}, \tag{6.9}$$

which describe the position errors and the orientation error between the actual robot position and direction and those of the virtual robot. It can be directly checked that these tracking errors satisfy the following differential equations:

$$\begin{bmatrix} \dot{x}_e \\ \dot{y}_e \\ \dot{\phi}_e \end{bmatrix} = \begin{bmatrix} wy_e + v - v_r\cos\phi_e \\ -wx_e + v_r\sin\phi_e \\ w - w_r \end{bmatrix}. \tag{6.10}$$

Introduce a new variable $\bar{\phi}_e = \phi_e + \arcsin(\dfrac{k(t)y_e}{\sqrt{1+x_e^2+y_e^2}})$ where

$$k(t) = \lambda v_r \tag{6.11}$$

with λ being a positive constant chosen so that $\lambda|v_r|^{\max} < 0.5$, where $|v_r|^{\max}$ denotes the maximum value of $|v_r(t)|$. Then (6.10) is transformed to

$$\begin{bmatrix} \dot{x}_e \\ \dot{y}_e \\ \dot{\bar{\phi}}_e \end{bmatrix} = \begin{bmatrix} wy_e + v + f_1 \\ -wx_e - \frac{k(t)v_r y_e}{\Gamma_1} + f_2 \\ w(1 - \frac{k(t)x_e}{\Gamma_1}) + f_3 - v\frac{k(t)x_e y_e}{\Gamma_1^2\Gamma_2} \end{bmatrix}, \tag{6.12}$$

where

$$\Gamma_1 = \sqrt{1 + x_e^2 + y_e^2}, \quad \Gamma_2 = \sqrt{1 + x_e^2 + (1 - k^2(t))y_e^2},$$

$$f_1 = -\frac{v_r \cos\bar{\phi}_e \Gamma_2}{\Gamma_1} - \frac{v_r k(t) y_e \sin\bar{\phi}_e}{\Gamma_1},$$

$$f_2 = -\frac{v_r(\cos\bar{\phi}_e - 1)\Gamma_2}{\Gamma_1} - \frac{v_r k(t) y_e \sin\bar{\phi}_e}{\Gamma_1},$$

$$f_3 = -w_r + \frac{\dot{k}(t)y_e + k(t)f_2 - k(t)(x_e f_1 + y_e f_2)/\Gamma_1^2}{\Gamma_2}. \tag{6.13}$$

Define virtual control errors $\tilde{v}$ and $\tilde{w}$ as

$$\tilde{v} = v - v_c, \quad \tilde{w} = w - w_c, \tag{6.14}$$

where v_c and w_c are the virtual controls to be designed for v and w. We have

$$\begin{bmatrix} \dot{x}_e \\ \dot{y}_e \\ \dot{\bar{\phi}}_e \end{bmatrix} = \begin{bmatrix} (w_c + \tilde{w})y_e + v_c + \tilde{v} + f_1 \\ -(w_c + \tilde{w})x_e - \frac{k(t)v_r y_e}{\Gamma_1} + f_2 \\ (w_c + \tilde{w})(1 - \frac{k(t)x_e}{\Gamma_1}) + f_3 - (v_c + \tilde{v})\frac{k(t)x_e y_e}{\Gamma_1^2 \Gamma_2} \end{bmatrix}. \tag{6.15}$$

Define a Lyapunov function

$$V_0 = \sqrt{1 + x_e^2 + y_e^2} + \sqrt{1 + \bar{\phi}_e^2} - 2.$$

The time derivative of V_0 is

$$\dot{V}_0 = -\frac{k(t)v_r y_e^2}{\Gamma_1^2} + \frac{x_e}{\Gamma_1}(v_c + f_1) + \frac{y_e f_2}{\Gamma_1} + \frac{\bar{\phi}_e}{\Gamma_3}[w_c(1-$$

$$\frac{k(t)x_e}{\Gamma_1}) + f_3 - v_c\frac{k(t)x_e y_e}{\Gamma_1^2 \Gamma_2}] + \tilde{v}(\frac{x_e}{\Gamma_1} - \frac{k(t)x_e y_e \bar{\phi}_e}{\Gamma_1^2 \Gamma_2 \Gamma_3})$$

$$+ \tilde{w}\frac{\bar{\phi}_e}{\Gamma_3}(1 - \frac{k(t)x_e}{\Gamma_1}).$$

The virtual control v_c and w_c are then chosen as

$$v_c = -k_1 \frac{x_e}{\Gamma_1} - f_1$$

$$w_c = \frac{1}{(1 - \frac{k(t)x_e}{\Gamma_1})}(-k_2\frac{\bar{\phi}_e}{\Gamma_3} - f_3 + v_c\frac{k(t)x_e y_e}{\Gamma_1^2 \Gamma_2}) \tag{6.16}$$

where $\Gamma_3 = \sqrt{1 + \bar{\phi}_e^2}$, and we get

$$\dot{V}_0 = -k_1\frac{x_e^2}{\Gamma_1^2} - \frac{k(t)v_r y_e^2}{\Gamma_1^2} - k_2\frac{\bar{\phi}_e^2}{\Gamma_3^2} + \frac{y_e f_2}{\Gamma_1}$$

$$+ \tilde{v}(\frac{x_e}{\Gamma_1} - \frac{k(t)x_e y_e \bar{\phi}_e}{\Gamma_1^2 \Gamma_2 \Gamma_3}) + \tilde{w}\frac{\bar{\phi}_e}{\Gamma_3}(1 - \frac{k(t)x_e}{\Gamma_1}).$$

Remark 6.1. Γ_1, Γ_2 and Γ_3 are introduced to normalize x_e, y_e and $\bar{\phi}_e$ so that the virtual control signals are globally bounded. This normalization technique will also be applied to the dynamic model when designing a control law for τ. The bounds of v_c and w_c are computable since $|\frac{x_e}{\Gamma_1}| < 1$, $|\frac{y_e}{\Gamma_1}| < 1$ and $|\frac{\bar{\phi}_e}{\Gamma_3}| < 1$. This will play an important role when proving the stability of the whole system, as will be shown later.

6.3.2 Design of State-feedback Control based on Dynamic Model

Define $\varpi_c = [v_c \ \ w_c]^T$, $\tilde{\varpi} = \varpi - \varpi_c$, and a new Lyapunov function

$$V_1 = V_0 + \frac{1}{2}\tilde{\varpi}^T R \tilde{\varpi}. \tag{6.17}$$

Then we get

$$\dot{V}_1 = \dot{V}_0 + \tilde{\varpi}^T R \dot{\tilde{\varpi}} = \dot{V}_0 + \tilde{\varpi}^T R(\dot{\varpi} - \dot{\varpi}_c)$$
$$= \dot{V}_0 + \tilde{\varpi}^T[-\bar{C}(w)\varpi - D_R\varpi + B_R\tau - R\dot{\varpi}_c]$$
$$= \dot{V}_0 + \tilde{\varpi}^T[-(\bar{C}(w) + D_R)\tilde{\varpi} + B_R\tau - Y_cH],$$

where $Y_cH = R\dot{\varpi}_c + \bar{C}(w)\varpi_c + D_R\varpi_c$ and

$$Y_c = \begin{bmatrix} \dot{v}_c & 0 & -w_cw & v_c & w_c & 0 \\ 0 & \dot{w}_c & v_cw & 0 & v_c & w_c \end{bmatrix}. \tag{6.18}$$

$$H = \begin{bmatrix} \frac{m_{11}+m_{22}}{b} & (m_{11}-m_{22})b & c & \frac{d_{11}+d_{22}}{2b} & \frac{d_{11}-d_{22}}{2} & \frac{b(d_{11}+d_{22})}{2} \end{bmatrix}^T. \tag{6.19}$$

Remark 6.2. By using v and w as virtual controls, the tracking error dynamic system in (6.12) and the virtual controllers v_c and w_c in (6.16) do not contain any unknown system parameters. So when taking the time-derivatives of v_c and w_c in the matrix Y_c in (6.18) to get the torque τ_i, no unknown parameter will appear. This will reduce the number of unknown parameters in (6.19) from 10 in [9] to 6 and thus greatly simplify the matrix Y_c. Therefore parameter estimation is simplified compared to [9].

Denote $\alpha := \frac{r}{2b}$, $\beta := \frac{r}{2}$ and $m = \frac{1}{\alpha}$, $n = \frac{1}{\beta}$. Let $\hat{m}$ and $\hat{n}$ be the estimates of m and n respectively, and define $\tilde{m} = m - \hat{m}$, $\tilde{n} = n - \hat{n}$, $u_1 = \frac{\tau_1 + \tau_2}{\hat{m}}$, $u_2 = \frac{\tau_1 - \tau_2}{\hat{n}}$, $U = [u_1 \ u_2]^T$. Then we get

$$B_R\tau = \begin{bmatrix} \alpha & \alpha \\ \beta & -\beta \end{bmatrix} \begin{bmatrix} \tau_1 \\ \tau_2 \end{bmatrix} = \begin{bmatrix} \alpha\hat{m}u_1 \\ \beta\hat{n}u_2 \end{bmatrix}. \tag{6.20}$$

As $\alpha\hat{m} = \alpha(m - \tilde{m}) = 1 - \alpha\tilde{m}$ and also $\beta\hat{n} = 1 - \beta\tilde{n}$,

$$B_R\tau = \begin{bmatrix} u_1 \\ u_2 \end{bmatrix} - \begin{bmatrix} \alpha\tilde{m}u_1 \\ \beta\tilde{n}u_2 \end{bmatrix} = U - \begin{bmatrix} \alpha\tilde{m}u_1 \\ \beta\tilde{n}u_2 \end{bmatrix}. \tag{6.21}$$

Let $\Omega_1 = (\dfrac{x_e}{\Gamma_1} - \dfrac{k(t)x_e y_e \bar{\phi}_e}{\Gamma_1{}^2 \Gamma_2 \Gamma_3})$, $\Omega_2 = \dfrac{\bar{\phi}_e}{\Gamma_3}(1 - \dfrac{k(t)x_e}{\Gamma_1})$, and

$$\Omega = [\Omega_1 \ \ \Omega_2]^T. \tag{6.22}$$

We now define a new Lyapunov function

$$V_2 = V_1 + \frac{1}{2}\tilde{H}^T \Lambda \tilde{H} + \frac{1}{2}\alpha\tilde{m}^2 + \frac{1}{2}\beta\tilde{n}^2 \tag{6.23}$$

where $\tilde{H} = H - \hat{H}$ and $\hat{H}$ is the estimate of H, $\Lambda = \gamma I$ is a positive definite matrix and γ is a positive constant. Then

$$
\begin{aligned}
\dot{V}_2 &= \dot{V}_0 + \tilde{\varpi}^T[-(\bar{C}(w) + D_R)\tilde{\varpi} + B_R\tau - Y_c H] + \tilde{H}^T \Lambda \dot{\tilde{H}} \\
&\quad + \alpha\tilde{m}\dot{\tilde{m}} + \beta\tilde{n}\dot{\tilde{n}} \\
&= -k_1\frac{x_e^2}{\Gamma_1^2} - \frac{k(t)v_r y_e^2}{\Gamma_1^2} - k_2\frac{\bar{\phi}_e^2}{\Gamma_3^2} + \frac{y_e f_2}{\Gamma_1} + \alpha\tilde{m}\dot{\tilde{m}} + \beta\tilde{n}\dot{\tilde{n}} \\
&\quad + \tilde{\varpi}^T[-(\bar{C}(w) + D_R)\tilde{\varpi} + B_R\tau - Y_c H + \Omega] + \tilde{H}^T \Lambda \dot{\tilde{H}} \\
&= -k_1\frac{x_e^2}{\Gamma_1^2} - \frac{k(t)v_r y_e^2}{\Gamma_1^2} - k_2\frac{\bar{\phi}_e^2}{\Gamma_3^2} + \frac{y_e f_2}{\Gamma_1} \\
&\quad - \alpha\tilde{m}(\dot{\hat{m}} + u_1\tilde{v}) - \beta\tilde{n}(\dot{\hat{n}} + u_2\tilde{w}) + \tilde{H}^T \Lambda \dot{\tilde{H}} \\
&\quad + \tilde{\varpi}^T[-(\bar{C}(w) + D_R)\tilde{\varpi} + U - Y_c H + \Omega].
\end{aligned}
\tag{6.24}
$$

By choosing U as

$$U = -K\tilde{\varpi} + Y_c\hat{H} - \Omega, \tag{6.25}$$

where $K = [\kappa_1 \ 0; 0 \ \kappa_2]$ with κ_1 and κ_2 being nonnegative constants, we get

$$
\begin{aligned}
\dot{V}_2 &= -k_1\frac{x_e^2}{\Gamma_1^2} - \frac{k(t)v_r y_e^2}{\Gamma_1^2} - k_2\frac{\bar{\phi}_e^2}{\Gamma_3^2} + \frac{y_e f_2}{\Gamma_1} - \tilde{\varpi}^T(D_R + K)\tilde{\varpi} \\
&\quad - \alpha\tilde{m}(\dot{\hat{m}} + u_1\tilde{v}) - \beta\tilde{n}(\dot{\hat{n}} + u_2\tilde{w}) - \tilde{H}^T T(\dot{\hat{H}} + \Lambda^{-1}Y_c^T \tilde{\varpi}).
\end{aligned}
$$

The control τ is designed as

$$\tau^{\text{state}} = \begin{bmatrix} \tau_1^{\text{state}} \\ \tau_2^{\text{state}} \end{bmatrix} = \begin{bmatrix} 1 & 1 \\ 1 & -1 \end{bmatrix}^{-1} \begin{bmatrix} \hat{m} & 0 \\ 0 & \hat{n} \end{bmatrix} U. \tag{6.26}$$

We use τ^{state} to signify it as the state feedback control in order to differentiate it from the output feedback control given later.

Now we design parameter estimators with a projection operation. Define $\theta = [m, n, H^T]^T$ and $g(v, w) = [-u_1\tilde{v}, -u_2\tilde{u}, -\Lambda^{-1}\tilde{\varpi}^T Y_c]^T$. Let $\hat{\theta}$ be the estimate of θ and $\tilde{\theta} = \theta - \hat{\theta}$. From Assumption 6.3, there exist $\theta_{i\max}$ and $\theta_{i\min}$ so that $\theta_{i\min} < \theta_i < \theta_{i\max}$. Define $\bar{\theta}_{i\min} = \frac{2-\sqrt{2}}{2}\theta_{i\min}$ and $\bar{\theta}_{i\max} = \frac{2+\sqrt{2}}{2}\theta_{i\max}$. The update law for $\hat{\theta}_i$, $i = 1, ..., 8$, is designed as follows:

$$\dot{\hat{\theta}}_{i\,state} = \begin{cases} g_i(v, w) & \text{if } \theta_{i\min} < \hat{\theta}_i < \theta_{i\max} \\ g_i(v, w) + (1 - \frac{\hat{\theta}_i}{\theta_{i\min}})^2[1 + g_i^2(v, w)] \\ \qquad \text{if } \hat{\theta}_i \leq \theta_{i\min} \\ g_i(v, w) - (1 - \frac{\hat{\theta}_i}{\theta_{i\max}})^2[1 + g_i^2(v, w)] \\ \qquad \text{if } \hat{\theta}_i \geq \theta_{i\max} \end{cases} \qquad (6.27)$$

where $g_i(v, w) = \vartheta_i^T g(v, w)$ and ϑ_i, $1 \leq i \leq 8$, is ith unit vector defined as

$$\vartheta_i = \begin{bmatrix} \overbrace{0...0}^{i-1} & 1 & \overbrace{0...0}^{8-i} \end{bmatrix}^T.$$

Here we use $\dot{\hat{\theta}}_{i\,state}$ also to signify it as a "state-feedback" estimator in which state variables are used.

Lemma 6.3. *The estimator (6.27) has the following property: (a) The parameter projection gives continuous $\dot{\hat{\theta}}_{i\,state}$. (b) Estimates $\hat{\theta}_i$ satisfies $\bar{\theta}_{i\min} \leq \hat{\theta}_i \leq \bar{\theta}_{i\max}$ for all $t \geq 0$ if $\bar{\theta}_{i\min} < \hat{\theta}_i(0) < \bar{\theta}_{i\max}$. (c) $-\alpha\tilde{m}(\dot{\hat{m}}+u_1\tilde{v}) - \beta\tilde{n}(\dot{\hat{n}}+u_2\tilde{w}) - \tilde{H}^T T(\dot{\hat{H}} + \Lambda^{-1} Y_c^T \tilde{\varpi}) \leq 0$ for all $t \geq 0$.*

Proof: (a) It is obvious that $\dot{\hat{\theta}}_{i\,state}$ is continuous. (b) When $\hat{\theta}_i = \bar{\theta}_{i\min}$, we have $(1 - \frac{\hat{\theta}_i}{\theta_{i\min}})^2 = \frac{1}{2}$, thus $\dot{\hat{\theta}}_i = \frac{1}{2}(1 + g_i(v, w))^2 \geq 0$ and this prevents $\hat{\theta}_i$ from getting smaller than $\bar{\theta}_{i\min}$. So $\hat{\theta}_i \geq \bar{\theta}_{i\min}$. Similarly, when $\hat{\theta}_i = \bar{\theta}_{i\max}$, we also have $(1 - \frac{\hat{\theta}_i}{\theta_{i\max}})^2 = \frac{1}{2}$ and thus $\dot{\hat{\theta}}_i = -\frac{1}{2}(1 + g_i(v, w))^2 < 0$ which prevents $\hat{\theta}_i$ from getting greater than $\bar{\theta}_{i\max}$. Therefore property (b) holds. (c) If $\hat{m} \leq m_{\min}$, then $\tilde{m} = m - \hat{m} \geq 0$ and $\dot{\hat{m}} + u_1\tilde{v} = (1 - \frac{\hat{m}}{m_{\min}})^2[1 + (u_1\tilde{v})^2] \geq 0$, therefore $-\tilde{m}(\dot{\hat{m}} + u_1\tilde{v}) \leq 0$. Similarly, when $\hat{m} \geq m_{\max}$, $\tilde{m} = m - \hat{m} \leq 0$ and $\dot{\hat{m}} + u_1\tilde{v} = -(1 - \frac{\hat{m}}{m_{\max}})^2[1 + (u_1\tilde{v})^2] \leq 0$. This ensures that $-\tilde{m}(\dot{\hat{m}} + u_1\tilde{v}) \leq 0$. When $m_{\min} < \hat{m} < m_{\max}$, $(\dot{\hat{m}} + u_1\tilde{v}) = 0$. We can also prove that $-\beta\tilde{n}(\dot{\hat{n}} + u_2\tilde{w}) - \tilde{H}^T T(\dot{\hat{H}} + \Lambda^{-1} Y_c^T \tilde{\varpi}) \leq 0$. Thus property (c) is established. $\square$

Based on Property (c) in Lemma 6.3, we have

$$\dot{V}_2 \leq -k_1\frac{x_e^2}{\Gamma_1^2} - \frac{k(t)v_r y_e^2}{\Gamma_1^2} - k_2\frac{\bar{\phi}_e^2}{\Gamma_3^2} + \frac{y_e f_2}{\Gamma_1}$$
$$- \tilde{\varpi}^T(D_R + K)\tilde{\varpi}. \qquad (6.28)$$

Proposition 6.4. *Under the adaptive controllers (6.22) and (6.25) - (6.27) with appropriate choices of parameters k_1, k_2, λ, δ_1 and δ_2, system (6.12) is uniformly stable and the trajectories of the virtual reference robot are asymptotically tracked.*

Proof: It is proved in Chapter 2 that positive constants $\delta_1 \in (0, 1)$ and $\delta_2 > 0$ can be chosen such that

$$\frac{y_e f_2}{\Gamma_1} \leq \frac{\delta_1 v_r{}^2 y_e{}^2}{\Gamma_1{}^2} + \delta_2 \frac{\bar{\phi}_e^2}{\Gamma_3^2}. \tag{6.29}$$

Substituting (6.29) into (6.28) yields

$$\dot{V}_2 \leq -k_1 \frac{x_e^2}{\Gamma_1^2} - \frac{(\lambda - \delta_1)v_r^2 y_e^2}{\Gamma_1^2} - (k_2 - \delta_2)\frac{\bar{\phi}_e^2}{\Gamma_3^2}$$
$$- \tilde{\varpi}^T(D_R + K)\tilde{\varpi}. \tag{6.30}$$

On the other hand, based on the condition that $|k(t)| < 0.5$, we choose positive constant λ to satisfy

$$\lambda v_r{}^{\max} < 0.5 \tag{6.31}$$

where $v_r{}^{\max}$ denotes the maximum value of $|v_r|$. From (6.30), the choices of λ and k_2 also need to meet the following conditions

$$\lambda > \delta_1 \quad \text{and} \quad k_2 > \delta_2. \tag{6.32}$$

Clearly the above choices of parameters are possible. Also we choose $k_1 > 0$. Let $k_3 = v_r{}^2(\lambda - \delta_1)$ and $k_4 = k_2 - \delta_2$, and we get

$$\dot{V}_2 \leq -\frac{k_1 x_e{}^2}{\Gamma_1{}^2} - \frac{k_3 y_e{}^2}{\Gamma_1{}^2} - \frac{k_4 \bar{\phi}_e{}^2}{\Gamma_3{}^2} - \tilde{\varpi}^T(D_R + K)\tilde{\varpi}. \tag{6.33}$$

Since $|v_r(t)| \geq r_\star$ for $t \geq T_s$, $k_3 > r_\star^2(\lambda - \delta_1)$ for $t \geq T_s$. Then from (6.33) we get $\lim\limits_{t \to \infty} |x_e| + |y_e| + |\bar{\phi}_e| + \| \tilde{\varpi} \| \to 0$.

Finally from the definition of $\bar{\phi}_e$, we know that $\lim\limits_{t \to \infty} y_e + |\bar{\phi}_e| \to 0$ guarantees $\lim\limits_{t \to \infty} |\phi_e| \to 0$. Therefore the trajectories are asymptotically tracked. $\square$

6.4 Adaptive Output Feedback Control

The design procedure is outlined as follows. Firstly two observers are designed to estimate the unknown states v and w respectively. Then the unknown states in the "state-feedback" control and parameter estimators presented in the previous section are replaced by the their estimates respectively. To prevent the peaking phenomenon of the high-gain observers from entering the robot dynamic system, we will invoke saturation to the controller and estimators. Finally system stability is analyzed and tracking errors are shown to remain in an arbitrarily small ball.

6.4.1 Observer Design

The main difficulty of adaptive output feedback control for nonholonomic mobile robots comes from the fact that designing state observers and parameter estimators are intertwined due to the presence of nonholonomic constraints and a quadratic term. In other words, the observers involve unknown parameters and the estimators involve unknown states as seen in (6.26) and (6.27). In this chapter we design two high-gain observers to estimate v and w respectively, which can overcome the difficulty. First, we define two new variables as follows

$$z_1 = y \cos \phi - x \sin \phi, \quad z_2 = x \cos \phi + y \sin \phi.$$

Then

$$\dot{z}_1 = -z_2 w$$
$$\dot{z}_2 = v + z_1 w$$
$$\dot{\phi} = w. \tag{6.34}$$

By noting the third equation of (6.34), following the standard high-gain observer design procedure in [58, 57], we design a high-gain observer for w as

$$\dot{\hat{\phi}} = \hat{w} + \frac{1}{\varepsilon} l_1 (\phi - \hat{\phi})$$
$$\dot{\hat{w}} = \frac{1}{\varepsilon^2} l_2 (\phi - \hat{\phi}), \tag{6.35}$$

where ε is a small positive parameter to be chosen, and l_i, $i = 1, 2$ are chosen so that

$$\bar{A} = \begin{bmatrix} -l_1 & 1 \\ -l_2 & 0 \end{bmatrix} \tag{6.36}$$

is Hurwitz. Noting the second equation of (6.34), the observer for v is designed as:

$$\dot{\hat{z}}_2 = \hat{v} + \frac{l_1}{\varepsilon} (z_2 - \hat{z}_2) + z_1 \hat{w}$$
$$\dot{\hat{v}} = \frac{l_2}{\varepsilon^2} (z_2 - \hat{z}_2), \tag{6.37}$$

where $\hat{w}$ is from system (6.35).

6.4.2 Controller and Estimator Design

To implement the output feedback control with high-gain observers, we need to introduce saturation limits to the control and the rates of the parameter estimates mainly for two reasons. The first is to prevent the peaking phenomena of the high-gain observers from entering the robot dynamic system. For a detailed description about peaking phenomena, please refer to [59]. The other is that the boundedness of the system signals is ensured by by saturation. On the other hand we will show that a short time after the operations of the observers and the estimators, the saturation will no longer take effect. This implies that the designed control signals and the rates of the parameter estimates will still remain within the saturation limits, independent of the presence of saturations.

We now define a compact set

$$\Xi = \{e : V_2 < \varrho\}, \tag{6.38}$$

where V_2 is defined in (6.23), $e = \{x_e, y_e, \bar{\phi}_e, \tilde{\varpi}, \tilde{H}, \tilde{m}, \tilde{n}\}$ and ρ is a constant defined as

$$\varrho = h_{\max} + \Delta_1,$$

where Δ_1 is a positive constant to be chosen such that $V_2(0) < \varrho$, and $h_{\max}$ is defined as

$$h_{\max} = \frac{1}{2}\tilde{H}_{\max}^T \Lambda \tilde{H}_{\max} + \frac{1}{2}\alpha \tilde{m}_{\max}^2 + \frac{1}{2}\beta \tilde{n}_{\max}^2$$

From Lemma 6.3, $\tilde{H}_i$, $1 \le i \le 6$, $\tilde{m}$ and $\tilde{n}$ are bounded. Denote $\tilde{H}_{i\max}$, $\tilde{m}_{\max}$, $\tilde{n}_{\max}$ as the maximum values of $\tilde{H}_i$, $\tilde{m}$ and $\tilde{n}$ respectively. Since τ_i^{state} in (6.26), $i = 1, 2$, is bounded and smooth in this compact set, let

$$S_i = \max_{e \subset \Xi} |\tau_i^{\text{state}}(\eta, \eta_r, \hat{\theta}, v, w)| + \Delta_2, \tag{6.39}$$

where $\eta = [x, y, \phi]^T$, $\eta_r = [x_r, y_r, \phi_r, v_r, w_r]^T$, and Δ_2 is a small positive constant to be chosen. From (6.16), (6.25), (6.26), we know S_i is computable.

The final saturated output feedback control inputs are designed as

$$\tau_i = \text{sat}_{S_i}\left(\tau_i^{\text{state}}(\eta, \eta_r, \hat{\theta}, \hat{v}, \hat{w})\right), \tag{6.40}$$

where the unknown v and w are replaced by $\hat{v}$ and $\hat{w}$ respectively and the saturation function is defined as

$$\text{sat}_{S_i}(\tau_i^{\text{state}}) = \begin{cases} \text{sign}(\tau_i^{\text{state}})S_i & |\tau_i^{\text{state}}| \ge S_i \\ \tau_i^{\text{state}} & |\tau_i^{\text{state}}| < S_i \end{cases} \tag{6.41}$$

Now we consider the design of estimators only involving output and observer signals. Define $\check{v} = \hat{v} - v_c$, $\check{w} = \hat{w} - w_c$, $\tilde{\varpi} = \hat{\varpi} - \varpi_c$. For the same reason as for the controllers, the rates of estimators will also be constrained by saturations in a similar way. The estimators for $\hat{\theta}_i$, $i = 1, ..., 8$, are designed as follows

$$\dot{\hat{\theta}}_i = \text{sat}_{S_{\hat{\theta}_i}}\left(g_{\hat{\theta}_i}(\hat{v}, \hat{m})\right), \tag{6.42}$$

where

$$g_{\hat{\theta}_i}(\hat{v}, \hat{m}) = \begin{cases} g_i(\hat{v}, \hat{w}) & \text{if } \theta_{i\,\min} < \hat{\theta}_i < \theta_{i\,\max} \\ g_i(\hat{v}, \hat{w}) + (1 - \frac{\hat{\theta}_i}{\theta_{i\,\min}})^2[1 + g_i^2(\hat{v}, \hat{w})] \\ \quad \text{if } \hat{\theta}_i \leq \theta_{i\,\min} \\ g_i(\hat{v}, \hat{w}) - (1 - \frac{\hat{\theta}_i}{\theta_{i\,\max}})^2[1 + g_i^2(\hat{v}, \hat{w})] \\ \quad \text{if } \hat{\theta}_i \geq \theta_{i\,\max} \end{cases}$$

$$S_{\hat{\theta}_i} = \max_{e \subset \Xi} |\dot{\hat{\theta}}_{istate}(v, w)| + \Delta_2.$$

$g_i(\hat{v}, \hat{m})$ is in the same form as $g_i(v, m)$ except that the unknown v and w are replaced by $\hat{v}$ and $\hat{w}$ respectively; $\text{sat}_{S_{\hat{\theta}_i}}(\cdot)$ is defined similarly to that of (6.41);

6.4.3 Stability Analysis

In this section we will give the stability analysis of the closed-loop system and show that the tracking errors are in an arbitrarily small ball eventually. This is roughly done in three steps. In the first step we prove that after a short time the saturation for controller and estimators will not take effect. Secondly we show that Ξ is an invariant set, which in turn implies the stability of the whole system. Finally we will establish that the tracking errors are in an arbitrarily small ball eventually.

Re-write (6.5) as follows

$$\dot{x} = v \cos\phi$$
$$\dot{y} = v \sin\phi$$
$$\dot{\phi} = w$$
$$\dot{v} = \alpha_1 w^2 + \alpha_2 v + \alpha_3 w + \alpha_4 \tau_1(\eta, \eta_r, \hat{\theta}, \hat{v}, \hat{w}) + \alpha_5 \tau_2(\eta, \eta_r, \hat{\theta}, \hat{v}, \hat{w})$$
$$\dot{w} = \alpha_6 vw + \alpha_7 v + \alpha_8 w + \alpha_9 \tau_1(\eta, \eta_r, \hat{\theta}, \hat{v}, \hat{w}) + \alpha_{10} \tau_2(\eta, \eta_r, \hat{\theta}, \hat{v}, \hat{w}), \quad (6.43)$$

where τ_1 and τ_2 are from (6.40), and a_i, $i = 1, ..., 10$ are unknown but bounded constants given as

$$\alpha_1 = -\frac{H(3)}{H(1)}, \alpha_2 = -\frac{H(4)}{H(1)}, \alpha_3 = -\frac{H(5)}{H(1)}, \alpha_4 = \alpha_5 = \frac{1}{mH(1)}$$
$$\alpha_6 = \frac{H(3)}{H(2)}, \alpha_7 = \frac{H(5)}{H(2)}, \alpha_8 = -\frac{H(6)}{H(2)}, \alpha_9 = -\alpha_{10} = \frac{1}{nH(2)}.$$

Define

$$\varphi_1 = \frac{1}{\varepsilon}(\phi - \hat{\phi}), \quad \varphi_2 = (w - \hat{w}) \qquad (6.44)$$

and denote $\varphi = [\varphi_1, \varphi_2]^T$. Then from (6.34) and the third and fifth equations of (6.43), we get

$$\varepsilon\dot{\varphi} = \bar{A}\varphi + \varepsilon \begin{bmatrix} 0 \\ G(v, w, \hat{v}, \hat{w}, \eta, \eta_r, \hat{\theta}) \end{bmatrix}, \tag{6.45}$$

where

$$G(v, w, \hat{v}, \hat{w}, \eta, \eta_r, \hat{\theta}) = \alpha_6 vw + \alpha_7 v + \alpha_8 w + \alpha_9 \tau_1 + \alpha_{10} \tau_2.$$

Define

$$\chi = [\chi_1 \ \chi_2]^T \tag{6.46}$$

where $\chi_1 = \dfrac{1}{\varepsilon}(z_2 - \hat{z}_2)$ and $\chi_2 = v - \hat{v}$. We also have

$$\varepsilon\dot{\chi} = \bar{A}\chi + \begin{bmatrix} z_1\varphi_2 \\ 0 \end{bmatrix} + \varepsilon \begin{bmatrix} 0 \\ H(v, w, \hat{v}, \hat{w}, \eta, \eta_r, \hat{\theta}) \end{bmatrix}, \tag{6.47}$$

where

$$H(v, w, \hat{v}, \hat{w}, \eta, \eta_r, \hat{\theta}) = \alpha_1 w^2 + \alpha_2 v + \alpha_3 w + \alpha_4 \tau_1 + \alpha_5 \tau_2. \tag{6.48}$$

Lemma 6.5. *When the system is controlled by the saturated controller (6.40) with estimator (6.42), φ defined in (6.44) is globally bounded.*

Proof: Consider a Lyapunov function $V_{\varpi} = \dfrac{1}{2}\varpi^T R\varpi$. Then from (6.2) and (6.8) we get

$$\dot{V}_{\varpi} \leq -2\frac{(\lambda_{\min}(D_R) - \frac{\varsigma}{2})}{\lambda_{\max}(R)}V_{\varpi} + \frac{1}{2\varsigma}\tau_b{}^T\tau_b$$

where $\tau_b = B_R\tau$ and ς is a constant chosen such that $\lambda_{\min}(D_R) - \frac{\varsigma}{2} > 0$. Since the torque τ_i, $i = 1, 2$ is saturated, so τ_b is also bounded. Thus v and w are globally bounded. Also from the boundedness of τ_i, there exists a positive constant g so that

$$|G(t, v, w, \hat{w}, \hat{\theta})| < g$$

Since $\bar{A}$ is a Hurwitz matrix, (6.45) is a stable system and φ is bounded. $\quad\square$

Lemma 6.6. *For any ε, there exists a finite time $T_1(\varepsilon)$, so that for $t > T_1$, $\| \varphi \| < c_1\varepsilon$ where c_1 is a positive constant that is computable.*

Proof: From **Lemma 6.5** we know φ is globally bounded. Define a positive Lyapunov function $\nu = \varphi^T P\varphi$ where matrix $P = [p_{11} \ p_{12}; p_{21} \ p_{22}]$ is the solution of the Lyapunov equation

$$\bar{A}^T P + P\bar{A} = -I$$

Then

$$\dot{\nu} \leq -\frac{1}{\varepsilon}\varphi^T\varphi + \varphi^T\left([(|p_{12}| + |p_{21}|)g \ \ 2|p_{22}|g]^T\right)$$
$$\leq -\frac{\gamma}{\varepsilon}\nu + G_3(|\varphi_1| + |\varphi_2|),$$

where $\gamma = 1/\lambda_{\max}(P)$ and $G_3 = \max\{g(|p_{12}| + |p_{21}|), 2g|p_{22}|\}$. It gives that

$$\dot{\nu} \leq -\frac{\gamma}{\varepsilon}\nu + \iota\sqrt{\nu}$$

where $\iota = \dfrac{\sqrt{2}G_3}{\sqrt{\lambda_{\min}(P)}}$. Now we consider two cases.

Case 1: For $\nu(t) \geq \dfrac{4\iota^2}{\gamma^2}\varepsilon^2$, we obtain

$$\dot{\nu} \leq -\frac{\gamma}{2\varepsilon}\nu \tag{6.49}$$

which results in that $\nu(t) \leq \nu(0)e^{-\frac{\gamma}{2\varepsilon}t}$. For any bounded initial conditions for ϕ, $\hat{\phi}$, w and $\hat{w}$, and from the definition of φ in (6.44), $\nu(0)$ can be written as $\nu(0) = \dfrac{l_3}{\varepsilon^2}$ where $l_3 = p_{11}s^2 + (p_{21} + p_{12})\varepsilon s q + p_{22}q^2\varepsilon^2$ with $s = \phi(0) - \hat{\phi}(0)$ and $q = w(0) - \hat{w}(0)$. Thus we get $\nu(t) \leq \dfrac{l_3}{\varepsilon^2}e^{-\frac{\gamma}{2\varepsilon}t}$.

So when the initial condition $\dfrac{l_3}{\varepsilon^2}$ satisfies that $\dfrac{l_3}{\varepsilon^2} > \dfrac{4\iota^2}{\gamma^2}\varepsilon^2$, the time $T_1(\varepsilon)$ for ν to reach the ball $\nu \leq \dfrac{4\iota^2}{\gamma^2}\varepsilon^2$ from $\nu(0)$ is

$$T_1(\varepsilon) = \frac{2\varepsilon}{\gamma}\left(\ln\frac{l_4}{\varepsilon^4}\right) \tag{6.50}$$

where $l_4 = \dfrac{l_3\gamma^2}{4\iota^2}$. Since (6.50) tends to zero as $\varepsilon \to 0$, so we can choose a sufficiently small $T_1(\varepsilon)$ such that $\forall t > T_1(\varepsilon)$, we have $\nu \leq \dfrac{4\iota^2}{\gamma^2}\varepsilon^2$, which means $\| \varphi \| \leq c_1\varepsilon$, where $c_1 = \sqrt{\dfrac{4\iota^2}{\gamma^2\lambda_{\min}(P)}}$.

Case 2: When $\nu(t) \leq \dfrac{4\iota^2}{\gamma^2}\varepsilon^2$, the conclusion holds automatically for any $T_1(\varepsilon) \geq 0$. □

Lemma 6.7. *There is a T_3 such that for $t \in [0, T_3)$, $e(t) \in \Xi$ if $e(0) \in \Xi$. Also there exists a $T_2(\varepsilon)$, which can be arbitrarily small if ε is small enough, such that for $t \in [T_2(\varepsilon), T_3)$, $\|\chi\| \leq c_2\varepsilon$.*

Proof: From (6.24) we get

$$\dot{V}_2 = \dot{V}_0 + \tilde{\varpi}^T[-(\bar{C}(w) + D_R)\tilde{\varpi} + B_R \tau - Y_c H]$$
$$+ \tilde{H}^T \Lambda \dot{\hat{H}} + \alpha \tilde{m}\dot{\hat{m}} + \beta \tilde{n}\dot{\hat{n}}. \tag{6.51}$$

With saturation implemented for the torque τ and the rates of estimates $\dot{\hat{H}}$, $\dot{\hat{m}}$, $\dot{\hat{n}}$, there is a positive constant c_4 such that $\dot{V}_2 < c_4$ when $V_2 \le \varrho$. Note that $V_2(t) = V_2(0) + \int_0^t \dot{V}_2 dt < V_2(0) + \int_0^t c_4 dt$ and $V_2(0) < \varrho$. Suppose T_3 is the first time that e exits Ξ, i.e. $V_2(T_3) = \varrho$. Then we get

$$T_3 > \frac{\varrho - V_2(0)}{c_4}$$

and the first part of the lemma is proved.

For $t \in (0, T_3)$, $|x_e| < \sqrt{\varrho^2 + 2\varrho}$ and $|y_e| < \sqrt{\varrho^2 + 2\varrho}$ from the definition of V_2. From (6.9), $|x| < |x_r| + |x_e| + |y_e|$ and $|y| < |y_r| + |x_e| + |y_e|$. Finally from Assumption 6.1 we get

$$|z_1| = |y\cos\phi - x\sin\phi| < |x| + |y| < 2a^\star + 4\sqrt{\varrho^2 + 2\varrho}.$$

Following the same proof as in Lemma 6.5 and Lemma 6.6, from (6.47) we know for $t \in (0, T_3)$, χ defined in (6.46) is bounded and there exist a computable positive constant c_2, and a $T_2(\varepsilon)$, which can be arbitrarily small if ε is small enough, such that for $t \in [T_2(\varepsilon), T_3)$, $\| \chi \| \le c_2 \varepsilon$. □

Let $T = \max\left(T_1(\varepsilon), T_2(\varepsilon)\right)$. From (6.50) we can find an $\varepsilon^\star$ to ensure

$$\max\left(T_1(\varepsilon^\star), T_2(\varepsilon^\star)\right) < T_3. \tag{6.52}$$

Then for all $\varepsilon < \varepsilon^\star$, it can be guaranteed that $T < T_3$.

Now we will show that the "output-feedback" parameter estimators also have those properties stated in Lemma 6.3 for $t \in [T, T_3)$.

Lemma 6.8. *For $t \in [T, T_3)$, the estimator (6.42) has the following properties: (a) $\dot{\hat{\theta}}_i$, $\dot{\hat{n}}$ and $\dot{\hat{H}}_i$ are all continuous. (b) Estimates $\hat{\theta}_i$ satisfy $\bar{\theta}_{i\min} \le \hat{\theta}_i \le \bar{\theta}_{i\max}$ for all $t \ge 0$ if $\bar{\theta}_{i\min} < \hat{\theta}_i(0) < \bar{\theta}_{i\max}$. (c) $-\tilde{m}(\dot{\hat{m}} + u_1\tilde{v}) - \tilde{n}(\dot{\hat{n}} + \hat{u}_2\tilde{w}) - \tilde{H}^T(\dot{\hat{H}} + \Lambda^{-1}Y_c^T\tilde{\varpi}) \le 0$ for all $t \ge 0$.*

Proof: As $\hat{\theta}_i(T) = \hat{\theta}_i(0) + \int_{t=0}^T \dot{\hat{\theta}}_i dt$ and the rates of the estimates are saturated, thus for $\bar{\theta}_{i\min} < \hat{\theta}_i(0) < \bar{\theta}_{i\max}$, an ε can be chosen to also guarantee

$$T \le \min_{i=1,\ldots,8}\left\{\frac{\min(\bar{\theta}_{i\max} - \hat{\theta}_i(0), \hat{\theta}_i(0) - \bar{\theta}_{i\min})}{|S_{\hat{\theta}_i}|}\right\}. \tag{6.53}$$

This will ensure

$$\bar{\theta}_{i\,\min} < \hat{\theta}_i(T) < \bar{\theta}_{i\,\max}. \tag{6.54}$$

Now from (6.25) and (6.26), there exist $\rho_{1i}(\eta, \eta_r, \hat{\theta})$ and $\rho_{2i}(\eta, \eta_r, \hat{\theta})$ such that

$$\tau_i^{\text{state}}(\eta, \eta_r, \hat{\theta}, v, w) = \bar{\tau}_i^{\text{state}}(\eta, \eta_r, \hat{\theta}) + \rho_{1i}v + \rho_{2i}w \tag{6.55}$$

where $\bar{\tau}_i^{\text{state}}(\eta, \eta_r, \hat{\theta})$ is the part that does not contain v and w. From (6.12), (6.16), (6.18), (6.25) and (6.26), $\rho_{1i}(t)$ and $\rho_{2i}(t)$ are bounded. Since x_e, y_e and $\bar{\phi}_e$ are normalized, it can also be checked that the bounds of ρ_{1i} and ρ_{2i} are computable. So $\tau_i^{\text{state}}(\eta, \eta_r, \hat{\theta}, v, w)$ and its output-feedback counterpart $\tau_i^{\text{state}}(\eta, \eta_r, \hat{\theta}, \hat{v}, \hat{w})$ have the following relationship for all $t \in [T, T_3)$

$$\begin{aligned}
\tau_i^{\text{state}}(\eta, \eta_r, \hat{\theta}, \hat{v}, \hat{w}) &= \bar{\tau}_i^{\text{state}}(\eta, \eta_r, \hat{\theta}) + \rho_{1i}\hat{v} + \rho_{2i}\hat{w} \\
&< \tau_i^{\text{state}}(\eta, \eta_r, \hat{\theta}) + \rho_{1i}v + \rho_{2i}w \\
&\quad + \rho_{1i}c_2\varepsilon + \rho_{2i}c_1\varepsilon \\
&< \tau_i^{\text{state}}(\eta, \eta_r, \hat{\theta}, v, w) + c_3\varepsilon
\end{aligned} \tag{6.56}$$

where c_3 is the bound of $\rho_{1i}c_2 + \rho_{2i}c_1$.

Note that parameter ε can be chosen so that

$$c_3\varepsilon < \Delta_2. \tag{6.57}$$

Thus for all $t \in [T, T_3)$, we have

$$\tau_i^{\text{state}}(\eta, \eta_r, \hat{\theta}, v, w) < \max_{e \in \Xi} |\tau_i^{\text{state}}(\eta, \eta_r, \hat{\theta}, v, w)|.$$

So from (6.39), (6.56) and (6.57) we know that during the interval $[T, T_3)$ the saturation will not take effect because $\tau_i^{\text{state}}(\eta, \eta_r, \hat{\theta}, \hat{v}, \hat{w}) < S_i$.

With the same analysis, no saturation will take effect for the rates of all the parameter estimates during this time interval either. Then by noting (6.54) and following similar analysis in the proof of Lemma 6.3, the same properties are established. □

Next we will show that T_3 in Lemma 6.7 is ∞; namely e will never exit Ξ and thus Ξ is an invariant set.

Lemma 6.9. *The set Ξ is an invariant set.*

Proof: From Lemma 6.7, (6.52) and (6.53), $e(t) \in \Xi$ for $t \in [0, T]$ if $e(0) \in \Xi$. Now defining $\tilde{\varpi} = [\tilde{v} \ \tilde{w}]^T$, we get $\dot{V}_2$ as follows

$$\begin{aligned}
\dot{V}_2 = &-k_1 \frac{x_e^2}{\Gamma_1^2} - \frac{k(t)v_r y_e^2}{\Gamma_1^2} - k_2 \frac{\bar{\phi}_e^2}{\Gamma_3^2} + \frac{y_e f_2}{\Gamma_1} - \tilde{\varpi}^T(D_R + K)\tilde{\varpi} \\
&- \alpha\tilde{m}(\dot{\hat{m}} + \hat{u}_1\tilde{v}) - \beta\tilde{n}(\dot{\hat{n}} + \hat{u}_2\tilde{w}) - \tilde{H}^T\Lambda(\dot{\hat{H}} \\
&+ \Lambda^{-1}\hat{Y}_c^T\tilde{\varpi}) + \alpha\tilde{m}[\hat{u}_1\tilde{v} - u_1\tilde{v}] + \beta\tilde{n}[\hat{u}_2\tilde{w} - u_2\tilde{w}] \\
&+ \tilde{H}^T[\hat{Y}_c^T\tilde{\varpi} - Y_c\tilde{\varpi}] + \tilde{\varpi}^T(\hat{Y}_c - Y_c)\hat{H}
\end{aligned}$$

for $t \in [T, T_3)$. According to Lemma 6.8, $\tilde{m}$, $\tilde{n}$, $\tilde{H}$, $\hat{m}$, $\hat{n}$ and $\hat{H}$ are bounded, and

$$-\alpha\tilde{m}(\dot{\hat{m}} + \hat{u}_1\tilde{v}) - \beta\tilde{n}(\dot{\hat{n}} + \hat{u}_2\tilde{w}) - \tilde{H}^T(\Lambda\dot{\hat{H}} + \hat{Y}_c^T\tilde{\varpi}) \leq 0.$$

Note that $\chi_2 = \tilde{v} - \check{v}$ and $\varphi_2 = \tilde{w} - \check{w}$, it is easy to check that $\hat{u}_1\tilde{v} - u_1\check{v}$, $\hat{u}_2\tilde{w} - u_2\check{w}$, $(\hat{Y}_c - Y_c)\hat{H}$ and $\hat{Y}_c^T\tilde{\varpi} - Y_c\check{\varpi}$ are polynomials of χ_2 and φ_2 of degree 2 without a constant term. Since $|\varphi_2| < c_1\varepsilon$ and $|\chi_2| < c_2\varepsilon$ from Lemmas 6.6 and 6.7 , we have

$$\dot{V}_2 \leq -\frac{k_1 x_e^2}{\Gamma_1^2} - \frac{k_3 y_e^2}{\Gamma_1^2} - \frac{k_4\bar{\phi}_e^2}{\Gamma_3^2} - \tilde{\varpi}^T(D_R + K)\tilde{\varpi} + k_6\varepsilon \tag{6.58}$$

for $t \in [T, T_3)$, where k_6 is a computable positive constant.

Define $\mathcal{E} = \{x_e, y_e, \bar{\phi}_e, \tilde{\varpi}\}$ and a new set

$$\Psi = \{\mathcal{E} : \frac{k_3 y_e^2}{\Gamma_1^2} + \frac{k_1 x_e^2}{\Gamma_1^2} + \frac{k_4\bar{\phi}_e^2}{\Gamma_3^2}$$
$$+ \tilde{\varpi}^T(D_R + K)\tilde{\varpi} \leq k_6\varepsilon\}. \tag{6.59}$$

Also denote ξ as

$$\xi = \max_{\mathcal{E}\in\Psi} V_1 \tag{6.60}$$

From (6.17) and (6.59), the design parameter ε can be also chosen to satisfy

$$\xi < \frac{1}{2}\Delta_1. \tag{6.61}$$

From the definitions of V_1 and Ψ this can be easily done. We now divide Ξ into two set Ξ_1 and Ξ_2 as follows

$$\Xi_1 = \{e : \mathcal{E} \in \Psi; \ e \in \Xi\} \quad \Xi_2 = \{e : \mathcal{E} \notin \Psi; \ e \in \Xi\}.$$

Clearly $\Xi = \Xi_1 \cup \Xi_2$ and $\Xi_1 \cap \Xi_2 = \emptyset$.

Ξ is divided so that when $e \in \Xi_1$, $V_2 < V_1 + h_{\max} < \frac{1}{2}\Delta_1 + h_{\max} = \varrho - \frac{1}{2}\Delta_1$, and when $e \in \Xi_2$, $\dot{V}_2 < 0$ from (6.58) and (6.59).

When $e \in \Xi_1$, $\dot{V}_2$ may be positive. But since $\dot{V}_2 < c_4$, $V_2(t)$ is continuous. Therefore if e exits Ξ_1, it will only enter Ξ_2, because $V_2(t) < \varrho - \frac{1}{2}\Delta_1$ when $e \in \Xi_1$. Thus when $\varrho - \frac{1}{2}\Delta_1 \leq V_2(t) < \varrho$, e is in Ξ_2. In another words, e will not exit Ξ from Ξ_1 directly. Ξ_1 and Ξ_2 are shown in Fig. 6.2 schematically.

Note that T_3 defined in Lemma 6.7 is the first time that e exits Ξ satisfying $V_2(T_3) = \varrho$. Then $V_2(t - \Delta t) < \varrho$, where $0 < \Delta t < T_3$ is an arbitrarily small constant. As stated, $e(t - \Delta t) \in \Xi_2$, so $\dot{V}_2(t - \Delta t) < 0$ and thus as Δt decreases $V_2(t - \Delta t)$ is monotonically decreasing. This contradicts that $V_2(T_3) = \varrho$. So

Fig. 6.2. Ξ is divided to Ξ_1 and Ξ_2.

e will not exit Ξ from Ξ_2 either, i.e. $T_3 = \infty$. Therefore Ξ is an invariant set.
$\square$

Since Ξ is an invariant set, then from Lemma 6.6 and Lemma 6.7,

$$\tau_i^{\text{state}}(\eta, \eta_r, \hat{\theta}, \hat{v}, \hat{w}) < S_i$$

for $t \in [T, \infty)$, which means the saturation will not take effect after T, and we have

$$\dot{V}_2 \leq -\frac{k_1 x_e^{\,2}}{\Gamma_1^{\,2}} - \frac{k_3 y_e^{\,2}}{\Gamma_1^{\,2}} - \frac{k_4 \bar{\phi}_e^{\,2}}{\Gamma_3^{\,2}} - \tilde{\varpi}^T (D_R + K)\tilde{\varpi} + k_6 \varepsilon \tag{6.62}$$

for $t \in [T, \infty)$. So based on Lemmas 6.5-6.9, we have the following theorem.

Theorem 6.10. *Consider the system consisting of a nonholonomic mobile robot (6.1) and (6.2), observers (6.35) and (6.37), the adaptive output feedback controllers τ in (6.40) and parameter estimators (6.42). Under Assumptions 6.1-6.3, and if $e(0) \in \Xi$, we can calculate an $\varepsilon^\star$ based on (6.52), (6.53), (6.57) and (6.61) so that for all $\varepsilon \in (0, \varepsilon^\star)$, (i) the signals in the systems are bounded; (ii) the tracking errors are of $O(\varepsilon)$; (iii) the steady state of tracking errors x_e, y_e, ϕ_e and $\tilde{\varpi}$ are within an arbitrarily small ball.*

Proof: (i) Since Ξ is an invariant set, thus all signals contained in Ξ are bounded. As shown in Lemma 6.6 and Lemma 6.7, φ and χ are bounded; thus all signals in the systems are bounded.
(ii) Define

$$\epsilon_1 = \frac{x_e}{\Gamma_1}, \epsilon_2 = \frac{y_e}{\Gamma_1}, \epsilon_3 = \frac{\bar{\phi}_e}{\Gamma_3} \tag{6.63}$$

and $\epsilon = [\epsilon_1, \epsilon_2, \epsilon_3, \tilde{\varpi}^T]^T$. Choose an $\varepsilon^\star$ satisfying (6.52), (6.53), (6.57) and (6.61), from (6.62) we have

$$\dot{V}_2 \leq -\epsilon^T Q \epsilon + k_6 \varepsilon \tag{6.64}$$

where

$$\bar{K} = \begin{bmatrix} k_1 & 0 & 0 \\ 0 & k_3 & 0 \\ 0 & 0 & k_4 \end{bmatrix}, \qquad Q = \begin{bmatrix} \bar{K} & 0 \\ 0 & D_R + K \end{bmatrix}.$$

Since V_2 is bounded and Q is positive definite, we have

$$\lim_{T \to \infty} \frac{1}{T} \int_0^T \epsilon^T \epsilon \leq k_7 \varepsilon \tag{6.65}$$

where $k_7 = k_6 / \lambda_{min}(Q)$, which shows that the mean-square of errors ϵ_1, ϵ_2, ϵ_3 and $\tilde{\varpi}$ are of the order of $O(\varepsilon)$ as in [58]. It can be further proved that the mean-square of tracking errors x_e, y_e, ϕ_e and $\tilde{\varpi}$ are also of the order of $O(\varepsilon)$.

(iii) Since $\dot{V}_2 \leq -\epsilon^T Q \epsilon + k_6 \varepsilon \leq -\lambda_{min}(Q)\epsilon^T \epsilon + k_6 \varepsilon$, so ϵ will finally reach the ball defined by

$$\epsilon^T \epsilon \leq k_7 \varepsilon. \tag{6.66}$$

From (6.64) we can see that x_e, y_e, $\bar{\phi}_e$, $\tilde{\varpi}$ will finally reach the ball defined by

$$x_e^2 + y_e^2 + \bar{\phi}_e^2 + \tilde{\varpi}^T \tilde{\varpi} \leq \left(\frac{2k_7}{1 - k_7 \varepsilon} + k_7 \right) \varepsilon \tag{6.67}$$

whose radius can be arbitrarily small since the design parameter ε can be arbitrarily small. Finally, note that $\phi_e = \bar{\phi}_e - \arcsin(\lambda v_r \epsilon_2)$. So ϕ_e is also in an arbitrarily small ball. □

Remark 6.11. For adaptive state-feedback tracking control of nonholonomic mobile robots, system stability is normally established under the persistent exciting (PE) condition, see for example, [1, 2, 9]. However, in order to achieve an extra tracking performance of small steady state tracking errors, in the sense that the tracking errors will eventually enter a given arbitrarily small ball and remain inside it, we need to guarantee $\lambda_{min}(Q) > \ell$ where ℓ is a positive constant. This is ensured by Assumption 6.2.

Remark 6.12. The keys for solving the adaptive output feedback control of the nonholonomic mobile robots lie in several aspects. First, observers for v and w are designed separately to avoid the nonholonomic constraint. Secondly saturation is introduced to the output feedback controller and estimators. This will prevent the peaking phenomenon from entering the robot system. Thirdly a compact set Ξ of e is set up. The stability of the whole system is guaranteed by proving Ξ is an invariant set.

Finally we summarize the overall design procedures in Table 6.1 and Table 6.2.

State-feedback controller:

$$U = -K\tilde{\varpi} + Y_c\hat{H} - \Omega$$

$$\tau^{\text{state}}(\eta, \eta_r, \hat{\theta}, v, w) = [\frac{1}{2}\hat{m} \quad \frac{1}{2}\hat{n}; \frac{1}{2}\hat{m} \quad -\frac{1}{2}\hat{n}]U$$

Observer:

$$\dot{\hat{\phi}} = \hat{w} + \frac{1}{\varepsilon}l_1(\phi - \hat{\phi})$$

$$\dot{\hat{w}} = \frac{1}{\varepsilon^2}l_2(\phi - \hat{\phi})$$

$$\dot{\hat{z}}_2 = \hat{v} + \frac{l_1}{\varepsilon}(z_2 - \hat{z}_2) + z_1\hat{w}$$

$$\dot{\hat{v}} = \frac{l_2}{\varepsilon^2}(z_2 - \hat{z}_2)$$

An invariant set:

$$\Xi = \{e : V_2 < \rho\}$$

Output feedback controller:

$$S_i = \max_{e \subset \Xi} |\tau_i^{\text{state}}(\eta, \eta_r, \hat{\theta}, v, w)| + \Delta_2$$

$$\tau_i = \text{sat}_{S_i}\left(\tau_i^{\text{state}}(\eta, \eta_r, \hat{\theta}, \hat{v}, \hat{w})\right)$$

Table 6.1. Summary of output feedback control for nonholonomic mobile robots.

6.5 Simulation Results

In this section, we present our simulation studies to illustrate our design scheme and verify the obtained theoretical results. The physical parameters of robots are given as: $b = 0.75$, $d = 0.3$, $r = 0.25$, $m_c = 10$, $m_w = 1$, $I_c = 5.6$, $I_w = 0.005$, $I_m = 0.0025$, $d_{11} = d_{22} = 1$. The initial estimates for the unknown parameters are set to be 60% of their true values. Based on the condition in *Proposition* 6.4, the following parameters are selected: $k_2 = 5$, $k_4 = 3$, $\lambda = 0.2$, $K_1 = 0$, $K_2 = 0$. From (6.52), (6.57) and (6.61), $\varepsilon^\star$ is calculated to be 0.021. We then choose Δ_1, Δ_2, ϱ, l_1, l_2, ε and Λ as 10, 1, 220, 10, 10, 0.001 and

Parameter estimator:

$$\dot{\hat{\theta}}_i = \text{sat}_{S_{\hat{\theta}_i}}\left(g_{\hat{\theta}_i}(\hat{v}, \hat{m})\right),$$

$$S_{\hat{\theta}_i} = \max_{e \subset \Xi} |\dot{\hat{\theta}}_{i\,state}(v, w)| + \Delta_2.$$

$$g_{\hat{\theta}_i}(\hat{v}, \hat{m}) = \begin{cases} g_i(\hat{v}, \hat{w}) & \text{if } \theta_{i\min} < \hat{\theta}_i < \theta_{i\max} \\ g_i(\hat{v}, \hat{w}) + (1 - \frac{\hat{\theta}_i}{\theta_{i\min}})^2[1 + g_i^2(\hat{v}, \hat{w})] \\ \qquad \text{if } \hat{\theta}_i \leq \theta_{i\min} \\ g_i(\hat{v}, \hat{w}) - (1 - \frac{\hat{\theta}_i}{\theta_{i\max}})^2[1 + g_i^2(\hat{v}, \hat{w})] \\ \qquad \text{if } \hat{\theta}_i \geq \theta_{i\max} \end{cases}$$

Table 6.2. Summary of parameter update laws.

$100I$ respectively. The saturation bound for τ_i, $\hat{m}$, $\hat{n}$ and $\hat{H}_i$ are chosen as 420, 10, 10 and 20 respectively. The maximum and minimum values for $\hat{m}$, $\hat{n}$, $\hat{H}(1)$ to $\hat{H}(6)$ are $\{(5, 20), (5, 20), (-2, 1), (-2, 1), (-2, 1), (5, 10), (-5, 5), (2, 5)\}$ respectively. The initial values for v, w and $\hat{v}$, $\hat{w}$ are set as $(0, 0)$. We use a circle as the reference. The initial position is $(2.3, 2)$, and the reference velocities are set as $v_r = 2.5$, $w_r = \dfrac{\pi}{2}$.

Simulation results are plotted in Fig.6.3-Fig.6.8. Fig.6.3 shows the tracking errors of the robot in (x, y) plane. Fig.6.4-Fig.6.6 show the time evolution of tracking errors x_e, y_e and ϕ_e respectively. It can be seen from the figures that the reference tracking errors are small, which is consistent with Theorem 6.10. Fig.6.7. shows the torque and it is observed that the saturation does not take effect after a short time. Fig.6.8 shows the estimates and the peaking phenomenon of the high-gain observers when estimating v and w. It can be seen that due to parameter projection the estimates are within the given bounds. Fig.6.10 shows the velocities of the two wheels.

6.6 Conclusions

In this chapter, a high-gain observer based adaptive output feedback tracking control design scheme for nonholonomic mobile robots is proposed. To overcome the difficulty due to the nonholonomic constraint and quadratic velocity term, we propose a new adaptive control scheme including designing a new adaptive state feedback controller and two high gain observers to estimate the unknown linear and angular velocities v and w respectively. The tracking errors can be confined to an arbitrarily small ball by choosing an appropriate design parameter ε. Simulation results validate obtained theoretical results. It should be noted that the observers can be extended to other underactuated mechanical systems, for example, underactuated ships.

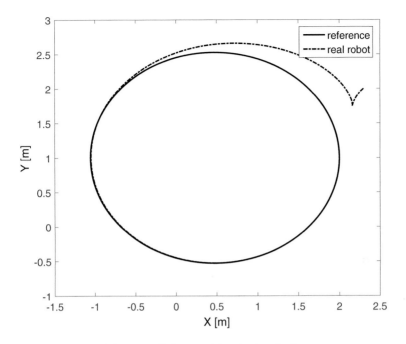

Fig. 6.3. Robot position in (x, y) plane.

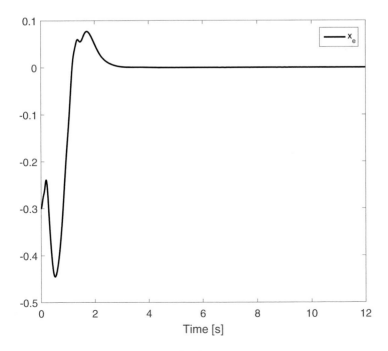

Fig. 6.4. Tracking error x_e.

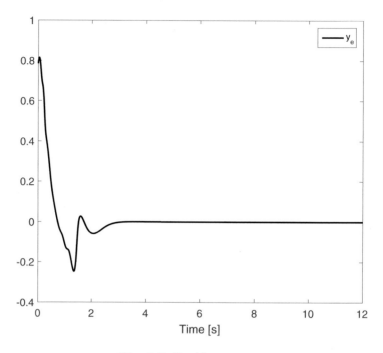

Fig. 6.5. Tracking error y_e.

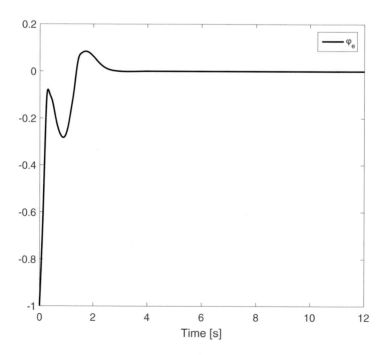

Fig. 6.6. Tracking error ϕ_e.

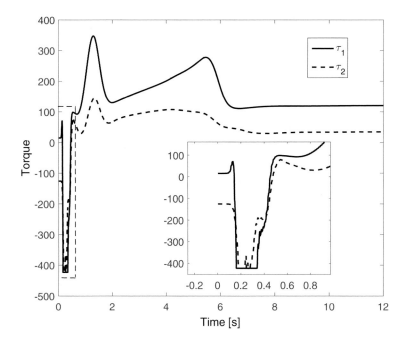

Fig. 6.7. Control torque τ_i, $i = 1, 2$.

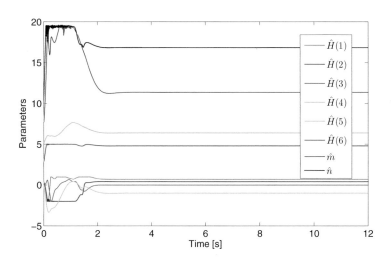

Fig. 6.8. The parameter estimators.

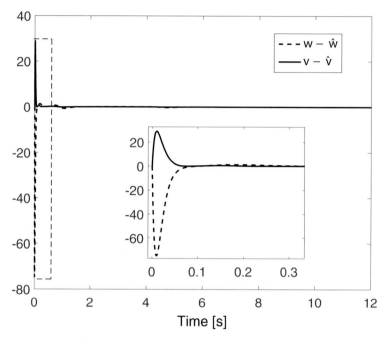

Fig. 6.9. Velocities estimate error and peaking phenomenon.

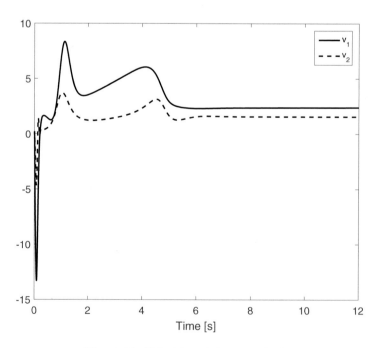

Fig. 6.10. Velocities of the two wheels.

7

Adaptive Output Feedback Control of an Underactuated Ship

In this chapter, both state feedback and output feedback robust adaptive controllers for an underactuated ship are proposed to make the ship follow a predefined path with guaranteed transient performance and arbitrarily small steady state position error and orientation error, despite the presences of environmental disturbances induced by wave, wind and ocean current and unknown system parameters. By using the prescribed performance bound technique, the position error and orientation error can be guaranteed converging to arbitrarily small residual sets at a pre-specified exponential rate. It is shown that the mean-square tracking errors are of order of an arbitrarily small design parameter, and the position error and the orientation error also decrease at a pre-specified exponential rate.

7.1 Introduction

The control of underactuated systems with nonholonomic constraints has received vast attention over the past few decades. Typical examples of this kind of systems include nonholonomic mobile robots, underactuated ships, underwater vehicles and VCTOL aircrafts, etc. Various control schemes are proposed for the path following control of underactuated ships. In [16] the method developed for chain-form systems was adopted for an underactuated ship through a coordinate transformation to steer both the position variable and the course angle of the ship, providing exponential stability for the tracking of the reference trajectory. In [60] by introducing a first-order sliding surface in terms of surge tracking errors and a second-order surface in terms of lateral motion tracking errors, a sliding-mode control law is proposed for trajectory tracking of underactuated ships. In [61] coordinated path-following controllers are designed for a group of underactuated ships to follow a reference path, where the derivative of each path is left as a free input to synchronize the ships' motions. In [62] an artificial immune system (AIS)-based adaptive control of generator excitation systems for electric ships with dramatic changes of disturbances is presented to solve power quality problems caused by high-energy loads. In [20] a continuous time-varying tracking controller was designed to yield globally uniformly ultimately bounded tracking by transforming the ship tracking system into a skew-symmetric form and designing a time-varying dynamic oscillator. In [42] a simple state-feedback control law was developed to render the tracking error globally $\mathcal{K}$-exponentially stable. In [21] a global robust adaptive controller was proposed to force an underactuated ship to follow a reference path under both constant and time-varying disturbances. In [24] a globally tracking controller was developed under the case that mass and damping matrices are not assumed to be diagonal and it was not required that the reference trajectory be generated by a ship model.

All the control schemes proposed above need the full information of the velocities. As most marine vessels only contain position sensors, the velocities in yaw, surge and sway directions are not available for feedback. This fact urges us to propose output feedback control schemes for the underactuated ships. In [25] through nonlinear coordinate changes to transform the ship dynamics to a system affine in the ship velocities, observers are designed to globally exponentially estimate unmeasured velocities. Related similar output feedback tracking control schemes are proposed in [54] for 1-DOF Lagrange systems or in [55] for multiple-DOF Lagrangian systems, where the observer design is based on a global nonlinear coordinate transformation to overcome the quadratic terms of the unmeasured states.

It should be pointed out that the observer design methods mentioned above need the full information of the system parameters. With unknown system parameters, several output feedback tracking control schemes are proposed. In [23] a nonlinear model-based adaptive output feedback controller was proposed for global asymptotic tracking in the presence of parametric uncertainties associated with nonlinear ship dynamics. In [63] and [31] a robust adaptive output feedback controller was designed for dynamic position of a surface vessel. However the ship model considered in these references is a fully-actuated surface ship. The main difficulty encountered lies in the simultaneous appearance of a nonholonomic constraint, unknown system parameters and the coriolis matrix, which results in quadratic cross terms involving unmeasured states. With this in mind, the observer design and the parameter estimator design are intertwined, which makes the analysis of the resulting closed loop system difficult and challenging. As far as we know, there is still no result in the literature concerning the control of underactuated ships when all system parameters are unknown and the yaw, surge and sway velocities are not available. Also the existing results guarantee that the tracking errors converge to zero only when time approaches infinity. So far there is no result considering transient performance. Note that, even though transient performance is difficult and a challenge to be addressed in adaptive control, it is an important specification in controller design that cannot be ignored. For instance, sometimes it is crucial for a ship to follow a prescribed path as soon as possible, to avoid the reef for example.

In this chapter, we address such issues by designing an adaptive controllers for the underactuated ships. Firstly, if all the states are available, we design adaptive state feedback controllers and state dependent parameter estimators to achieve path following control. Secondly when the yaw, surge and sway velocities are unavailable, we design three high-gain observers to estimate these velocities. Thirdly, we replace the states in the state-feedback controllers and the parameter estimators with the estimated velocities from the observers. Due to their ability to reject disturbances, it is shown that the estimation errors generated by the high-gain observers are of order $O(\epsilon)$, where ϵ is a design parameter. This in turn enables the tracking errors of an underactuated ship, including the position error and the orientation error, to be made arbitrarily small. Our proposed controller is shown performing effectively in tracking control of underactuated ships by simulation studies. Also by incorporating the prescribed performance bound technique [65], the designed controllers guarantee that the position error and the orientation error converge to arbitrarily small values with convergence rates faster than any pre-specified exponential speed, that is a specification for transient performance.

7.2 Problem Formulation

7.2.1 System Model and Control Objective

The motion of the ship can be described by the following differential equations, see for example[21]:

$$\begin{bmatrix} \dot{x} \\ \dot{y} \\ \dot{\phi} \end{bmatrix} = \begin{bmatrix} \cos\phi & -\sin\phi & 0 \\ \sin\phi & \cos\phi & 0 \\ 0 & 0 & 1 \end{bmatrix} \begin{bmatrix} u \\ v \\ r \end{bmatrix}, \tag{7.1}$$

$$\dot{u} = \frac{m_{22}}{m_{11}}vr - \frac{d_{11}}{m_{11}}u - \sum_{i=2}^{3}\frac{d_{ui}}{m_{11}}|u|^{i-1}u + \frac{1}{m_{11}}\tau_u + \frac{1}{m_{11}}\tau_{wu}(t)$$

$$\dot{v} = -\frac{m_{11}}{m_{22}}ur - \frac{d_{22}}{m_{22}}v - \sum_{i=2}^{3}\frac{d_{vi}}{m_{22}}|v|^{i-1}v + \frac{1}{m_{22}}\tau_{wv}(t)$$

$$\dot{r} = \frac{m_{11}-m_{22}}{m_{33}}uv - \frac{d_{33}}{m_{33}}r - \sum_{i=2}^{3}\frac{d_{ri}}{m_{33}}|r|^{i-1}r + \frac{1}{m_{33}}\tau_r + \frac{1}{m_{33}}\tau_{wr}(t) \tag{7.2}$$

where (x, y) denotes the coordinates of the surface ship, ϕ is the heading angle of the ship; u, v and r denote the velocity in surge, sway and yaw respectively. τ_u and τ_r are surge force and yaw force respectively, which are considered as inputs. Since the sway force is not available, the system (7.1) and (7.2) are underactuated. The positive constants m_{jj} denote the ship inertia including

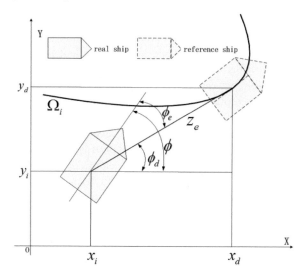

Fig. 7.1. General framework of ship following a path.

adding mass, and d_{jj}, d_{ui}, d_{vi}, d_{ri}, $2 \leq i \leq 3$, $1 \leq j \leq 3$ represent the hydrodynamic damping coefficients in surge, sway and yaw respectively. These constants are unknown. The bounded time-varying terms, $\tau_{wu}(t)$, $\tau_{wv}(t)$ and $\tau_{wr}(t)$, are the environmental disturbances induced by the wave, wind and ocean-current. We assume that

$$|\tau_{wu}(t)| < \tau_{wu\,max}, \quad |\tau_{wv}(t)| < \tau_{wv\,max}, \quad |\tau_{wr}(t)| < \tau_{wr\,max} \qquad (7.3)$$

where $\tau_{wu\,max}$, $\tau_{wv\,max}$ and $\tau_{wr\,max}$ are unknown positive constants.

Control Objective: For system (7.1) and (7.2), design controls τ_u and τ_r to force the above mentioned ship to follow a prescribed path set Ω as denoted by $(x_d(s),\ y_d(s))$ such that the position error and the orientation error will converge to arbitrarily small values with a pre-specified exponential convergence rate, where s is a path parameter as shown in Fig.7.1. Path $(x_d(s), y_d(s))$ may be regarded as generated by a "reference ship".

The control objective is to be achieved with both state feedback and output feedback controls subject to the following assumptions.

Assumption 7.1 $x'_d(s)^2 + y'_d(s)^2 \geq \mu$ *where* μ *is a positive constant,* $x'_d(s) = \dfrac{\partial x_d(s)}{\partial s}$ *and* $y'_d(s) = \dfrac{\partial y_d(s)}{\partial s}$. *The minimum radius of the osculating circle of the path is larger than or equal to the minimum possible turning radius of the ships.*

Assumption 7.2 *The unknown parameters are in known convex compact sets.*

Assumption 7.3 $|x_d(s)| < a^\star, |y_d(s)| < a^\star$, *where* $a^\star$ *is a known but arbitrary positive constant.*

Remark 7.1.
• Assumption 7.1 implies that the path Ω is regular with respect to the path parameter s. If the path is not regular, it can be divided into several different regular paths for consideration separately.
• Theoretically $a^\star$ in Assumption 7.3 could be arbitrarily large to cover any practical reference path. Normally a commercial path is composed by several point-to-point segments where each segment has a limited distance, which can be governed by $a^\star$.

We now define the following error variables [21]:

$$x_e = x_d - x, \quad y_e = y_d - y,$$
$$\phi_e = \phi - \phi_d, \quad z_e = \sqrt{x_e^2 + y_e^2}, \tag{7.4}$$

where $\phi_d = arcsin\dfrac{y_e}{z_e}$, and variables ϕ_e and z_e are shown in Figure 7.1. Since the system modeled in (7.1) and (7.2) is underactuated, this brings difficulty to controller design. To overcome this, adaptive control schemes will be proposed to ensure that z_e and ϕ_e converge to arbitrarily small values with convergence rates faster than a pre-specified exponential speed. In addition, the dynamics in the sway direction will also be stabilized. Clearly these enable our control objective to be achieved.

From (7.4), we obtain that

$$\sin(\phi_e) = \frac{x_e \sin(\phi) - y_e \cos(\phi)}{z_e}, \quad \cos(\phi_e) = \frac{x_e \cos(\phi) + y_e \sin(\phi)}{z_e}. \tag{7.5}$$

From (7.4), (7.5), and (7.1), the derivatives of z_e, ϕ_e are computed as

$$\dot{z}_e = -\cos(\phi_e)u + \sin(\phi_e)v + \left(\frac{x_e}{z_e}x_d' + \frac{y_e}{z_e}y_d'\right)\dot{s}$$

$$\dot{\phi}_e = r + \left(\frac{\sin(\phi)}{z_e \cos(\phi_e)} - \frac{x_e \tan(\phi_e)}{z_e^2}\right)(x_d'\dot{s} - u\cos(\phi) + v\sin(\phi))$$

$$- \left(\frac{\cos(\phi)}{z_e \cos(\phi_e)} + \frac{y_e \tan(\phi_e)}{z_e^2}\right)(y_d'\dot{s} - u\sin(\phi) - v\cos(\phi)) \tag{7.6}$$

where $x_d' = \dfrac{\partial x_d}{\partial s}$ and $y_d' = \dfrac{\partial y_d}{\partial s}$.

7.2.2 Performance Characterization and System Transformation

From (7.6), it can be seen that the system will be un-defined if $z_e = 0$ and $\phi_e = \pm\dfrac{\pi}{2}$. To avoid this, the control inputs need be designed such that $|\phi_e| < \dfrac{\pi}{2}$ and $z_e > \sigma_1$ holds $\forall 0 \leq t \leq \infty$ with σ_1 being a positive constant.

Based on this consideration, our proposed control scheme needs to ensure the following inequalities hold.

$$\sigma_1 < z_e(t) \leq \delta, \quad -\frac{\pi}{2} + \sigma_2 \leq \phi_e(t) \leq \frac{\pi}{2} - \sigma_2 \tag{7.7}$$

where δ and $\sigma_2 < \dfrac{\pi}{2}$ are positive constants. Thus an upper bound and a lower bound for the position error and orientation error are prescribed in addition to stability and steady state tracking properties. Such constraints bring new difficulties to our controller design. To overcome them, the prescribed performance bounds (PPB) approach in [65] is incorporated. We first introduce a smooth performance function $\eta(t)$ as follows

$$\eta(t) = (\eta_0 - \eta_\infty)e^{-at} + \eta_\infty, \tag{7.8}$$

where $a > 0$ and $\eta_0 \geq \eta_\infty > 0$ which are all chosen by users. Note that if $\eta_0 > \eta_\infty$, $\eta(t)$ is an exponentially decreasing function with a as the decaying rate and $\lim_{t\to\infty} \eta(t) = \eta_\infty$. Next by satisfying the following conditions,

$$\alpha_1\eta(t) < z_e(t) < \beta_1\eta(t), \quad -\alpha_2\eta(t) < \phi_e(t) < \beta_2\eta(t) \tag{7.9}$$

with positive scalars α_1, β_1, α_2 and β_2 chosen as

$$\alpha_1 \geq \frac{\sigma_1}{\eta_\infty}, \quad \beta_1 \leq \frac{\delta}{\eta_0}, \tag{7.10}$$

$$\alpha_2 \leq \frac{\pi - 2\sigma_2}{2\eta_0}, \quad \beta_2 \leq \frac{\pi - 2\sigma_2}{2\eta_0} \tag{7.11}$$

ensuring "constrained" errors in (7.7) is achieved. Moreover by converting (7.7) to (7.9), the transient performances of tracking errors with a as a bound of exponentially decaying rate is also guaranteed by choosing $\eta_0 > \eta_\infty$. Thus $\eta(t)$ characterizes both the transient and steady-state performances of the closed loop system. It is worthy pointing out that all the existing results on underactuated ship adaptive control do not consider transient performances, to the authors' best knowledge.

However, it is still a challenging problem to ensure (7.9) for all $t \geq 0$. To overcome the difficulty, we transform it to solving a problem with boundedness of certain signals as the only requirements. To do this, design two smooth and strictly increasing functions $S_1(\chi)$ and $S_2(\gamma)$ with the following properties:

$$(i) \quad \alpha_1 < S_1(\chi) < \beta_1, \quad -\alpha_2 < S_2(\gamma) < \beta_2 \tag{7.12}$$

$$(ii) \quad \lim_{\chi \to +\infty} S_1(\chi) = \beta_1, \quad \lim_{\chi \to -\infty} S_1(\chi) = \alpha_1$$

$$\lim_{\gamma \to +\infty} S_2(\gamma) = \beta_2, \quad \lim_{\gamma \to -\infty} S_2(\gamma) = -\alpha_2 \tag{7.13}$$

$$(iii) \quad S_1(0) = \alpha_1 + \epsilon < \beta_1, \quad S_2(0) = 0 \tag{7.14}$$

where ϵ is a positive constant chosen sufficiently small. From properties (i) and (ii) of S_1 and S_2, the "constrained" error conditions in (7.9) can be expressed as

$$z_e(t) = \eta(t)S_1(\chi), \quad \phi_e(t) = \eta(t)S_2(\gamma). \tag{7.15}$$

Because of the strict monotonicity property of S_1 and S_2, and the fact that $\eta(t) \neq 0$, the inverse functions exist and are given as

$$\chi(t) = S_1^{-1}\left(\frac{z_e(t)}{\eta(t)}\right), \quad \gamma(t) = S_2^{-1}\left(\frac{\phi_e(t)}{\eta(t)}\right). \tag{7.16}$$

If $\alpha_1\eta(0) < z_e(0) < \beta_1\eta(0)$, $-\alpha_2\eta(0) < \phi_e(0) < \beta_2\eta(0)$, and the transformed errors $\chi(t)$ and $\gamma(t)$ are ensured bounded $\forall t \geq 0$ by our proposed adaptive controllers, we will have that $\alpha_1 < \dfrac{z_e(t)}{\eta(t)} \leq \beta_1$ and $-\alpha_2 < \dfrac{\phi_e(t)}{\eta(t)} \leq \beta_2$, and thus (7.9) is ensured. Furthermore, from property (iii), if $\lim_{t\to\infty} \gamma(t) = 0$ and $\lim_{t\to\infty} \chi(t) = 0$ are obtained, we get $\lim_{t\to\infty} \phi_e(t) = 0$ and $\lim_{t\to\infty} z_e(t) = \eta_\infty(\alpha_1 + \epsilon) > \eta_\infty\alpha_1 \geq \sigma_1$. One possible choice of $S_1(\chi)$ and $S_2(\gamma)$ is

$$S_1(\chi) = \frac{\beta_1 e^{(\chi+v_1)} + \alpha_1 e^{-(\chi+v_1)}}{e^{(\chi+v_1)} + e^{-(\chi+v_1)}}, \quad S_2(\gamma) = \frac{\beta_2 e^{(\gamma+v_2)} - \alpha_2 e^{-(\gamma+v_2)}}{e^{(\gamma+v_2)} + e^{-(\gamma+v_2)}} \tag{7.17}$$

where $v_1 = \dfrac{1}{2}\ln\dfrac{\epsilon}{\beta_1 - \alpha_1 - \epsilon}$, $v_2 = \dfrac{1}{2}\ln\dfrac{\alpha_2}{\beta_2}$. It can be easily shown that $S_1(\chi)$ and $S_2(\gamma)$ in (7.17) have the required properties (i)-(iii). From (7.16), the transformed errors χ and γ are then solved as

$$\chi = S_1^{-1}(\lambda_1(t)) = \frac{1}{2}\ln[\lambda_1(t) - \alpha_1] + \frac{1}{2}\ln(\beta_1 - \alpha_1 - \epsilon) - \frac{1}{2}\ln[\epsilon\beta_1 - \epsilon\lambda_1(t)]$$

$$\gamma = S_2^{-1}(\lambda_2(t)) = \frac{1}{2}\ln\left(\lambda_2(t)\beta_2 + \alpha_2\beta_2\right) - \frac{1}{2}\ln\left(\alpha_2\beta_2 - \lambda_2(t)\alpha_2\right) \tag{7.18}$$

where $\lambda_1(t) = \dfrac{z_e(t)}{\eta(t)}$ and $\lambda_2(t) = \dfrac{\phi_e(t)}{\eta(t)}$. Taking the time-derivative of χ and γ, respectively, we obtain

$$\dot{\chi} = \frac{\partial S_1^{-1}}{\partial \lambda_1}\dot{\lambda}_1 = \frac{1}{2}\left[\frac{1}{\lambda_1(t) - \alpha_1} - \frac{1}{\lambda_1(t) - \beta_1}\right]\left(\frac{\dot{z}_e}{\eta} - \frac{z_e}{\eta^2}\dot{\eta}\right),$$

$$\dot{\gamma} = \frac{\partial S_2^{-1}}{\partial \lambda_2}\dot{\lambda}_2 = \frac{1}{2}\left[\frac{1}{\lambda_2(t) + \alpha_2} - \frac{1}{\lambda_2(t) - \beta_2}\right]\left(\frac{\dot{\phi}_e}{\eta} - \frac{\phi_e}{\eta^2}\dot{\eta}\right). \tag{7.19}$$

Replacing $\dot{z}_e$ and $\dot{\phi}_e$ with $\dot{\chi}$ and $\dot{\gamma}$, the system dynamics are transformed to

$$\dot{\chi} = \xi_1\left[-\cos(\phi_e)u + \sin(\phi_e)v + \left(\frac{x_e}{z_e}x_d' + \frac{y_e}{z_e}y_d'\right)\dot{s} - \frac{z_e\dot{\eta}}{\eta}\right]$$

$$\dot{\gamma} = \xi_2\left[r + \left(\frac{\sin(\phi)}{z_e\cos(\phi_e)} - \frac{x_e\tan(\phi_e)}{z_e^2}\right)(x_d'\dot{s} - u\cos(\phi) + v\sin(\phi))\right.$$

$$\left. - \left(\frac{\cos(\phi)}{z_e\cos(\phi_e)} + \frac{y_e\tan(\phi_e)}{z_e^2}\right)(y_d'\dot{s} - u\sin(\phi) - v\cos(\phi)) - \frac{\phi_e\dot{\eta}}{\eta}\right]$$

$$\dot{u} = \frac{m_{22}}{m_{11}}vr - \frac{d_{11}}{m_{11}}u - \sum_{i=2}^{3}\frac{d_{ui}}{m_{11}}|u|^{i-1}u + \frac{1}{m_{11}}\tau_u + \frac{1}{m_{11}}\tau_{wu}(t)$$

$$\dot{v} = -\frac{m_{11}}{m_{22}}ur - \frac{d_{22}}{m_{22}}v - \sum_{i=2}^{3}\frac{d_{vi}}{m_{22}}|v|^{i-1}v + \frac{1}{m_{22}}\tau_{wv}(t)$$

$$\dot{r} = \frac{m_{11}-m_{22}}{m_{33}}uv - \frac{d_{33}}{m_{33}}r - \sum_{i=2}^{3}\frac{d_{ri}}{m_{33}}|r|^{i-1}r + \frac{1}{m_{33}}\tau_r + \frac{1}{m_{33}}\tau_{wr}(t) \quad (7.20)$$

where $\xi_1 = \frac{1}{2\eta}\left[\frac{1}{\lambda_1(t)-\alpha_1} - \frac{1}{\lambda_1(t)-\beta_1}\right]$ and $\xi_2 = \frac{1}{2\eta}\left[\frac{1}{\lambda_2(t)+\alpha_2} - \frac{1}{\lambda_2(t)-\beta_2}\right]$.
By applying property (i) of $S_1(\chi)$ and $S_2(\gamma)$, it can also be easily checked that $\xi_1 \neq 0$ and $\xi_2 \neq 0$.

7.3 Adaptive State-feedback Control Design

Under the general framework of backstepping technique [8], we shall design the control input τ_u and τ_r to ensure stability of the transformed system (7.20) in two steps. In the first step, virtual controls u_d and r_d for u and r can be designed to guarantee the stability of χ and γ. In the second step real controls τ_u and τ_r will be designed. To address the unknown system parameters, suitable parameter estimators will be introduced.

• Step 1.

Define two virtual control error variables

$$\tilde{u} = u - u_d, \quad \tilde{r} = r - r_d \quad (7.21)$$

and consider the following Lyapunov function

$$V_1(t) = \frac{1}{2}\chi^2 + \frac{1}{2}\gamma^2. \quad (7.22)$$

By considering the time-derivative of (7.22), the virtual controls u_d and r_d are designed as

$$u_d = \frac{k_1}{\xi_1 \cos(\phi_e)}\chi + \frac{1}{\cos(\phi_e)}\left[\sin(\phi_e)v + \left(\frac{x_e}{z_e}x'_d + \frac{y_e}{z_e}y'_d\right)\dot{s} - \frac{z_e\dot{\eta}}{\eta}\right]$$

$$r_d = -\frac{k_2}{\xi_2}\gamma - \left(\frac{\sin(\phi)}{z_e\cos(\phi_e)} - \frac{x_e\tan(\phi_e)}{z_e^2}\right)(x'_d\dot{s} - u_d\cos(\phi) + v\sin(\phi))$$

$$+ \left(\frac{\cos(\phi)}{z_e\cos(\phi_e)} + \frac{y_e\tan(\phi_e)}{z_e^2}\right)(y'_d\dot{s} - u_d\sin(\phi) - v\cos(\phi)) + \frac{\phi_e\dot{\eta}}{\eta} \quad (7.23)$$

where k_1 and k_2 are positive constants, and $\dot{s}$ is selected to be

$$\dot{s} = \frac{u_0(1 - \epsilon_* e^{-\epsilon^* t})}{\sqrt{x'^2_{od} + y'^2_{od}}} \tag{7.24}$$

with u_0, $\epsilon_* < 1$ and ϵ^* being positive constants. Then the derivative of V_1 becomes

$$\dot{V}_1 = -k_1\chi^2 - k_2\gamma^2 - \chi\xi_1\cos(\phi_e)\tilde{u} + \xi_2\gamma\tilde{r} \tag{7.25}$$

$$-\xi_2\gamma\Big(\cos(\phi)f_1 - \sin(\phi)f_2\Big)\tilde{u} \tag{7.26}$$

where $f_1 = \dfrac{\sin(\phi)}{z_e\cos(\phi_e)} - \dfrac{x_e\tan(\phi_e)}{z_e^2}$ and $f_2 = \dfrac{\cos(\phi)}{z_e\cos(\phi_e)} + \dfrac{y_e\tan(\phi_e)}{z_e^2}$.

Remark 7.2. Note that similar to the existing references [21], $\dot{s}$ is closely related to the speed of the ship, which is determined to guarantee the feasibility of reference trajectory to be followed by a real ship. That is, the "reference ship" runs at a low speed in the beginning and speeds up slowly. Moreover, when a sudden turn is encountered, i.e. $\sqrt{x'^2_{od} + y'^2_{od}}$ becomes larger, $\dot{s}$ turns smaller, which means the real ship will slow down. So compared to the tracking control, the existence of $\dot{s}$ makes a path more practical for a real ship to follow.

- *Step 2.* Taking the derivatives of $\tilde{u}$ and $\tilde{r}$ along (7.21) gives

$$\dot{\tilde{u}} = \frac{m_{22}}{m_{11}}vr - \frac{d_{11}}{m_{11}}u - \sum_{i=2}^{3}\frac{d_{ui}}{m_{11}}|u|^{i-1}u + \frac{1}{m_{11}}\tau_u + \frac{1}{m_{11}}\tau_{wu}(t)$$

$$- \frac{\partial u_d}{\partial \xi_1}\dot{\xi}_1 - \frac{\partial u_d}{\partial \phi_e}\dot{\phi}_e - \frac{\partial u_d}{\partial \chi}\dot{\chi} - \frac{\partial u_d}{\partial x_e}\dot{x}_e - \frac{\partial u_d}{\partial y_e}\dot{y}_e - \frac{\partial u_d}{\partial z_e}\dot{z}_e - \frac{\partial u_d}{\partial x'_d}x''_d$$

$$- \frac{\partial u_d}{\partial y'_d}y''_d - \frac{\partial u_d}{\partial \dot{s}}\ddot{s} - \frac{\partial u_d}{\partial \eta}\dot{\eta} - \frac{\partial u_d}{\partial \dot{\eta}}\ddot{\eta} - \frac{\partial u_d}{\partial v}\Big(-\frac{m_{11}}{m_{22}}ur$$

$$- \frac{d_{22}}{m_{22}}v - \sum_{i=2}^{3}\frac{d_{vi}}{m_{22}}|v|^{i-1}v + \frac{1}{m_{22}}\tau_{wv}(t)\Big)$$

$$\dot{\tilde{r}} = \frac{m_{11} - m_{22}}{m_{33}}uv - \frac{d_{33}}{m_{33}}r - \sum_{i=2}^{3}\frac{d_{ri}}{m_{33}}|r|^{i-1}r + \frac{1}{m_{33}}\tau_r + \frac{1}{m_{33}}\tau_{wr}(t)$$

$$- \frac{\partial r_d}{\partial \xi_2}\dot{\xi}_2 - \frac{\partial r_d}{\partial \phi_e}\dot{\phi}_e - \frac{\partial r_d}{\partial \gamma}\dot{\gamma} - \frac{\partial r_d}{\partial x_e}\dot{x}_e - \frac{\partial r_d}{\partial y_e}\dot{y}_e - \frac{\partial r_d}{\partial z_e}\dot{z}_e - \frac{\partial r_d}{\partial x'_d}x''_d$$

$$- \frac{\partial r_d}{\partial y'_d}y''_d - \frac{\partial r_d}{\partial \dot{s}}\ddot{s} - \frac{\partial r_d}{\partial \eta}\dot{\eta} - \frac{\partial r_d}{\partial \dot{\eta}}\ddot{\eta} - \frac{\partial r_d}{\partial v}\Big(-\frac{m_{11}}{m_{22}}ur$$

$$- \frac{d_{22}}{m_{22}}v - \sum_{i=2}^{3}\frac{d_{vi}}{m_{22}}|v|^{i-1}v + \frac{1}{m_{22}}\tau_{wv}(t)\Big). \tag{7.27}$$

We define the Lyapunov function at this step as

$$V_2 = V_1 + \frac{m_{11}}{2}\tilde{u}^2 + \frac{m_{33}}{2}\tilde{r}^2 + \sum_{i=1}^{3} \tilde{\theta}_i^T \Gamma_i^{-1}\tilde{\theta}_i \qquad (7.28)$$

where $\tilde{\theta}_i = \theta_i - \hat{\theta}_i$ is the parameter estimation error, $\Gamma_1 = diag\{\delta_{1j}\}$, $\Gamma_2 = diag\{\delta_{2j}\}$, $j = 1,...,9$, $\Gamma_3 = diag\{\delta_{3k}\}$, $k = 1,...,4$, are positive definite diagonal matrices , and the unknown parameter vector θ_i, $i = 1, 2, 3$ is defined as

$$\theta_1 = \left[m_{22},\, d_{11},\, d_{u2},\, d_{u3},\, m_{11},\, \frac{m_{11}^2}{m_{22}},\, \frac{m_{11}d_{22}}{m_{22}},\, \frac{m_{11}d_{v2}}{m_{22}},\, \frac{m_{11}d_{v3}}{m_{22}}\right]$$

$$\theta_2 = \left[(m_{11} - m_{22}),\, d_{33},\, d_{v2},\, d_{v3},\, m_{33},\, \frac{m_{11}m_{33}}{m_{22}},\, \frac{m_{33}d_{22}}{m_{22}},\, \frac{m_{33}d_{v2}}{m_{22}},\, \frac{m_{33}d_{v3}}{m_{22}}\right]$$

$$\theta_3 = \left[\tau_{wu\,max},\, \frac{m_{11}}{m_{22}}\tau_{wv\,max},\, \tau_{wr\,max},\, \frac{m_{33}}{m_{22}}\tau_{wv\,max}\right]. \qquad (7.29)$$

Without canceling the useful damping term, the control torques for τ_u and τ_r are designed as

$$\tau_{us} = -k_3\tilde{u} - \hat{\theta}_1^T \Omega_1 + \chi\xi_1\cos(\phi_e) + \xi_2\gamma\left(\cos(\phi)f_1 - \sin(\phi)f_2\right)$$

$$- \hat{\theta}_{31}\tanh\left(\frac{\tilde{u}\hat{\theta}_{31}}{\epsilon_1}\right) - \hat{\theta}_{32}\frac{\partial u_d}{\partial v}\tanh\left(\frac{\partial u_d}{\partial v}\frac{\tilde{u}\hat{\theta}_{32}}{\epsilon_2}\right)$$

$$\tau_{rs} = -k_4\tilde{r} - \hat{\theta}_2^T \Omega_2 - \xi_2\gamma - \hat{\theta}_{33}\tanh\left(\frac{\tilde{r}\hat{\theta}_{33}}{\epsilon_3}\right) - \hat{\theta}_{34}\frac{\partial r_d}{\partial v}\tanh\left(\frac{\partial r_d}{\partial v}\frac{\tilde{r}\hat{\theta}_{34}}{\epsilon_4}\right) \quad (7.30)$$

where k_3 and k_4 are positive constants, and

$$\Omega_1 = \left[vr,\, -u_d,\, -|u|u_d,\, -u^2u_d,\, -\left(\frac{\partial u_d}{\partial\xi_1}\dot{\xi}_1\right.\right.$$

$$+ \frac{\partial u_d}{\partial\phi_e}\dot{\phi}_e + \frac{\partial u_d}{\partial\chi}\dot{\chi} + \frac{\partial u_d}{\partial x_e}\dot{x}_e + \frac{\partial u_d}{\partial y_e}\dot{y}_e + \frac{\partial u_d}{\partial z_e}\dot{z}_e$$

$$\left.+ \frac{\partial u_d}{\partial x_d'}x_d'' + \frac{\partial u_d}{\partial y_d'}y_d'' + \frac{\partial u_d}{\partial\dot{s}}\ddot{s} + \frac{\partial u_d}{\partial\eta}\dot{\eta} + \frac{\partial u_d}{\partial\dot{\eta}}\ddot{\eta}\right),$$

$$\left.\frac{\partial u_d}{\partial v}ur,\, \frac{\partial u_d}{\partial v}v,\, \frac{\partial u_d}{\partial v}|v|v,\, \frac{\partial u_d}{\partial v}v^3\right]$$

$$\Omega_2 = \left[uv,\, -r_d,\, -|r|r_d,\, -r^2r_d,\, -\left(\frac{\partial r_d}{\partial\xi_2}\dot{\xi}_2\right.\right.$$

$$+ \frac{\partial r_d}{\partial\phi_e}\dot{\phi}_e + \frac{\partial r_d}{\partial\gamma}\dot{\gamma} + \frac{\partial r_d}{\partial x_e}\dot{x}_e + \frac{\partial r_d}{\partial y_e}\dot{y}_e + \frac{\partial r_d}{\partial z_e}\dot{z}_e$$

$$\left.+ \frac{\partial r_d}{\partial x_d'}x_d'' + \frac{\partial r_d}{\partial y_d'}y_d'' + \frac{\partial r_d}{\partial\dot{s}}\ddot{s} + \frac{\partial r_d}{\partial\eta}\dot{\eta} + \frac{\partial r_d}{\partial\dot{\eta}}\ddot{\eta}\right),$$

$$\left.\frac{\partial r_d}{\partial v}ur,\, \frac{\partial r_d}{\partial v}v,\, \frac{\partial r_d}{\partial v}|v|v,\, \frac{\partial r_d}{\partial v}v^3\right].$$

Here τ_{us} and τ_{rs} are used to denote that they are the state-feedback controllers, differentiating them from the output-feedback controls τ_u and τ_r, which will be designed later.

The update law for $\hat{\theta}_i$ are chosen as

$$\dot{\hat{\theta}}_{1js} = \delta_{1j}\mathrm{Proj}(\tilde{u}\Omega_{1j}, \hat{\theta}_{1j}), \quad 1 \le j \le 9$$

$$\dot{\hat{\theta}}_{2js} = \delta_{2j}\mathrm{Proj}(\tilde{r}\Omega_{2j}, \hat{\theta}_{2j}), \quad 1 \le j \le 9$$

$$\dot{\hat{\theta}}_{31s} = \delta_{31}\mathrm{Proj}(|\tilde{u}|, \hat{\theta}_{31})$$

$$\dot{\hat{\theta}}_{32s} = \delta_{32}\mathrm{Proj}\left(\left|\tilde{u}\frac{\partial u_d}{\partial v}\right|, \hat{\theta}_{32}\right)$$

$$\dot{\hat{\theta}}_{33s} = \delta_{33}\mathrm{Proj}(|\tilde{r}|, \hat{\theta}_{33})$$

$$\dot{\hat{\theta}}_{34s} = \delta_{34}\mathrm{Proj}\left(\left|\tilde{r}\frac{\partial r_d}{\partial v}\right|, \hat{\theta}_{34}\right). \tag{7.31}$$

Similarly $\dot{\hat{\theta}}_{**s}$ is used to denote it as a "state-feedback" estimator in which unknown state variables are used. The operator Proj is a Lipschitz continuous projection algorithm [94] defined as follows:

$$\mathrm{Proj}(a, \hat{b}) = \begin{cases} a & \text{if } \mu(\hat{b}) \le 0 \\ a & \text{if } \mu(\hat{b}) \ge 0 \text{ and } \mu'(\hat{b})a \le 0 \\ (1 - \mu(\hat{b}))a & \text{if } \mu(\hat{b}) > 0 \text{ and } \mu'(\hat{b})a > 0 \end{cases} \tag{7.32}$$

where $\mu(\hat{b}) = \dfrac{\hat{b}^2 - b_M^2}{\epsilon^2 + 2\epsilon b_M}$, $\mu'(\hat{b}) = \dfrac{\partial\mu(\hat{b})}{\hat{b}}$, ϵ is an arbitrarily small positive constant, b_M is a positive constant satisfying $|b| < b_M$. Based on [64], for any $\epsilon > 0$ and $\zeta \in \Re$,

$$0 \le |\zeta| - \zeta\tanh\left(\frac{\zeta}{\epsilon}\right) \le 0.2785\epsilon. \tag{7.33}$$

Differentiating $V_2(t)$ and using (7.30), (7.31), (7.33) and Lemma 7.12 yields that

$$\dot{V}_2 \le -k_1\chi^2 - k_2\gamma^2 - (k_3 + d_{11})\tilde{u}^2 - (k_4 + d_{33})\tilde{r}^2 + 0.2785\sum_{i=1}^{4}\epsilon_i. \tag{7.34}$$

By adding and subtracting $\frac{1}{2}\sum_{i=1}^{3}\tilde{\theta}_i^T\Gamma_i^{-1}\tilde{\theta}_i$ on the right side of (7.34), we get

$$\dot{V}_2 \le -k_m V_2 + \epsilon_m \tag{7.35}$$

where $k_m = \min\left(2k_1, 2k_2, \dfrac{2(k_3 + d_{11})}{m_{11}}, \dfrac{2(k_4 + d_{33})}{m_{33}}\right)$, $\epsilon_m = \frac{1}{2}\sum_{i=1}^{3}\tilde{\theta}_i^T\Gamma_i^{-1}\tilde{\theta}_i + 0.2785\sum_{i=1}^{4}\epsilon_i$. From (7.35), it is directly shown that

$$V_2(t) \le V_2(0)e^{-k_m t} + \frac{\epsilon_m}{k_m} \tag{7.36}$$

which means χ, γ, $\tilde{u}$, $\tilde{r}$, $\tilde{\theta}_i$, $i = 1, 2, 3$ are bounded. Thus from (7.15) z_e and ϕ_e are bounded, and also (7.7) will be satisfied.

Let

$$u_d = a_1 v + a_2, \quad r_d = a_3 v + a_4 \tag{7.37}$$

where $a_1 = \dfrac{\sin(\phi_e)}{\cos(\phi_e)}$, $a_2 = \dfrac{k_1}{\xi_1 \cos(\phi_e)} \chi + \dfrac{1}{\cos(\phi_e)} \left[\left(\dfrac{x_e}{z_e} x'_d + \dfrac{y_e}{z_e} y'_d \right) \dot{s} - \dfrac{z_e \dot{\eta}}{\eta} \right]$,

$a_3 = f_1 a_1 \cos(\phi) - f_1 \sin(\phi) - f_2 a_1 \sin(\phi) - f_2 \cos(\phi)$ and $a_4 = -\dfrac{k_2}{\xi_2} \gamma - $

$f_1 (x'_d \dot{s} - a_2 \cos(\phi)) + f_2 (y'_d \dot{s} - a_2 \sin(\phi)) + \dfrac{\phi_e \dot{\eta}}{\eta}$. From (7.36) a_1, a_2, a_3 and a_4 are bounded. The sway velocity dynamics can be rewritten as

$$
\begin{aligned}
\dot{v} &= -\frac{m_{11}}{m_{22}}(a_1 v + a_2 + \tilde{u})(a_3 v + a_4 + \tilde{r}) - \frac{d_{22}}{m_{22}} v \\
&\quad - \sum_{i=2}^{3} \frac{d_{vi}}{m_{22}} |v|^{i-1} v + \frac{1}{m_{22}} \tau_{wv}(t) \\
&= a_5(t) v + a_6(t) v^2 - \frac{d_{v2}}{m_{22}} |v| v - \frac{d_{v3}}{m_{33}} v^3 + a_7(t)
\end{aligned}
\tag{7.38}
$$

where $a_5(t) = -\dfrac{m_{11}}{m_{22}} \left(a_1(a_4 + \tilde{r}) + a_3(a_2 + \tilde{u}) \right) - \dfrac{d_{22}}{m_{22}}$, $a_6(t) = -\dfrac{m_{11}}{m_{22}} a_1 a_3$,

$a_7(t) = -\dfrac{m_{11}}{m_{22}} (a_2 + \tilde{u})(a_4 + \tilde{r}) + \dfrac{1}{m_{22}} \tau_{wv}(t)$. From the fact that χ, γ, $\tilde{u}$, $\tilde{r}$, a_1, a_2, a_3 and a_4 are bounded, it can be easily checked that a_5, a_6 and a_7 are bounded. Also from Assumption 7.2, (7.7) and (7.36), the upper-bounds of a_5, a_6 and a_7 can also be obtained. Considering the time-derivative of Lyapunov function $V_v = \dfrac{1}{2} v^2$ along (7.38), we get

$$\dot{V}_v \leq -\left(\frac{d_{v3}}{m_{22}} - a_6^M \zeta \right) v^4 + \left(a_5^M + \frac{a_6^M + a_7^M}{4\zeta} \right) v^2 + a_7^M \zeta \tag{7.39}$$

where ζ is a positive constant chosen such that $\dfrac{d_{v3}}{m_{22}} - a_6^M \zeta > 0$. a_i^M, $i = 5, 6, 7$ are the upper-bounds of a_i respectively. Then from [66] it can be guaranteed that v is bounded with computable upper-bound. The following theorem is thus established.

Theorem 7.3. *Consider the underactuated ship system (7.1) and (7.2), with the state-feedback controller (7.30) and parameter update laws (7.31). Under Assumptions 7.1 and 7.2, all the signals in the closed-loop dynamic system are bounded, and the reference path $x_d(s)$ and $y_d(s)$ can be followed with arbitrarily small position error and orientation error. Furthermore, the position error and orientation error will converge at a pre-specified exponential rate.*

Proof: From (7.36), the boundedness of χ, γ, $\tilde{u}$, $\tilde{r}$, $\tilde{\theta}_{1i}$, $\tilde{\theta}_{2i}$, $i = 1, ..., 9$, $\tilde{\theta}_{3j}$, $j = 1, ..., 4$ and V_2 are shown. From (7.38) v is bounded, thus from (7.23) u_d and r_d are bounded. Then u and r are also bounded. It is easy to check from (7.30) and (7.31) that τ_{us}, τ_{rs}, $\dot{\hat{\theta}}_{1s}$, $\dot{\hat{\theta}}_{2s}$ and $\dot{\hat{\theta}}_{3s}$ are bounded. Let

$$e = [\chi, \ \gamma, \ \tilde{u}, \ \tilde{r}]^T \tag{7.40}$$

and $k_{min} = \min\{k_1, \ k_2, \ k_3 + d_{11}, \ k_4 + d_{33}\}$. Taking the integration of (7.34) yields

$$\lim_{T \to \infty} \frac{1}{T} \int_{t=0}^{T} e^T e \le \frac{0.2785 \sum_{i=1}^{4} \epsilon_i}{k_{min}}$$

which means e is of the order of $O(\epsilon_i)$, i.e., e can be adjusted arbitrarily small if ϵ_i is small enough. From (7.15), the properties of $S_1(\chi)$, $S_2(\gamma)$ and (7.11), we obtain that

$$\sigma_1 \le \alpha_1 \eta_\infty < z_e(t) < \beta_1 \eta_0 \le \delta$$
$$-\frac{\pi}{2} + \sigma_2 \le -\alpha_2 \eta_0 < \phi_e(t) < \beta_2 \eta_0 \le \frac{\pi}{2} - \sigma_2, \quad \forall \ t \ge 0. \tag{7.41}$$

if $z_e(0) \in \left(\alpha_1 \eta_\infty, \beta_1 \eta_0\right)$ and $\phi_e(0) \in (-\pi/2 + \sigma_2, \pi/2 - \sigma_2)$, then $\lim_{t \to \infty} \phi_e(t) = \epsilon_\star$ and $\lim_{t \to \infty} z_e(t) = (\alpha_1 + \epsilon + \epsilon^\star)\eta_\infty \in \left(\alpha_1 \eta_\infty, \beta_1 \eta_\infty\right)$ where $\epsilon_\star$ and $\epsilon^\star$ are unknown constants but are arbitrarily small due to the continuity of $S_1(\chi)$ and $S_2(\gamma)$ and the fact that χ and γ are arbitrarily small. It implies that (7.7) is always satisfied, and the underactuated ship can follow its reference trajectory with arbitrary small position error z_e and orientation error ϕ_e.
From (7.9)

$$z_e(t) < \beta_1 \left((\eta_0 - \eta_\infty)e^{-at} + \eta_\infty\right), \quad \phi_e(t) < \beta_2 \left((\eta_0 - \eta_\infty)e^{-at} + \eta_\infty\right) \tag{7.42}$$

which means that the position error and the orientation error will converge at a rate governed by the given positive constant a. This completes the proof. $\square$

7.4 Adaptive Output-feedback Control Design

In this section, we consider the case that all the velocities are unavailable for feedback. As a starting point, three high-gain observers are designed to estimate the unknown velocities respectively. Then we replace the unknown states in the state feedback controllers and state dependent parameter estimators with the estimated velocities. Finally, the closed loop system is analyzed analysis is to show its stability and tracking properties.

7.4.1 Observer Design

The difficulty of adaptive output feedback control for underactuated ships comes from the fact that the observer design and the unknown parameters are intertwined together due to the presence of nonholonomic constraints and a quadratic term of the system states. Now we design three high-gain observers to estimate u, v and r respectively, which can avoid the difficulty mentioned above. Firstly, we make the following variable changes

$$\rho_1 = y\cos\phi - x\sin\phi, \quad \rho_2 = x\cos\phi + y\sin\phi$$

Then from (7.1)

$$\dot{\rho}_1 = v - \rho_2 r$$
$$\dot{\rho}_2 = u + \rho_1 r$$
$$\dot{\phi} = r. \tag{7.43}$$

Following the procedure in [58] and based on the third equation of (7.43), we design a high-gain observer for r as

$$\dot{\hat{\phi}} = \hat{r} + \frac{1}{\varepsilon_1}l_1(\phi - \hat{\phi})$$

$$\dot{\hat{r}} = \frac{1}{\varepsilon_1^2}l_2(\phi - \hat{\phi}) \tag{7.44}$$

where ε_1 is a small positive design parameter, and l_i, $i = 1, 2$ are chosen so that

$$\bar{A} = \begin{bmatrix} -l_1 & 1 \\ -l_2 & 0 \end{bmatrix} \tag{7.45}$$

is Hurwitz. From the second equation of (7.43), the observer for u is designed as:

$$\dot{\hat{\rho}}_2 = \hat{u} + \frac{l_1}{\varepsilon_1}(\rho_2 - \hat{\rho}_2) + \rho_1\hat{r}$$

$$\dot{\hat{u}} = \frac{l_2}{\varepsilon_1^2}(\rho_2 - \hat{\rho}_2) \tag{7.46}$$

where $\hat{r}$ is from system (7.44).
From the first equation of (7.43), the observer for v is designed as:

$$\dot{\hat{\rho}}_1 = \hat{v} + \frac{l_1}{\varepsilon_1}(\rho_1 - \hat{\rho}_1) - \rho_2\hat{r}$$

$$\dot{\hat{v}} = \frac{l_2}{\varepsilon_1^2}(\rho_1 - \hat{\rho}_1). \tag{7.47}$$

7.4.2 Design of Adaptive Controllers and Estimators

Since the output feedback controllers and the parameter estimators are implemented with the high-gain observers, we will introduce saturation to the controllers and parameter estimators based on two reasons. The first reason is to prevent the peaking phenomena of the high-gain observers from entering the ship dynamic system. The other is that the boundedness of the system signals is ensured by saturation. On the other hand we will show that after a short time from the beginning of system operation, the saturation will have no effect. This implies that the designed torques and the rates of the parameter estimates will still remain within the saturation limits, independent of the presence of saturations.

First of all, we define a compact set

$$\Xi = \{\mathcal{E} : V_2 < \varrho\} \tag{7.48}$$

where V_2 is defined in (7.28) and $\mathcal{E} = \{\tilde{u}, \tilde{r}, \chi, \gamma, \tilde{\theta}_{1i}, \tilde{\theta}_{2i}, \tilde{\theta}_{3j}\}$, $i = 1, ..., 9$, $j = 1, ..., 4$. Due to Assumption 7.2 and the parameter projections, $\tilde{\theta}_{1i}$, $\tilde{\theta}_{2i}$, $\tilde{\theta}_{3j}$ are bounded with computable upper-bounds. Denote $\tilde{\theta}_{1im}$, $\tilde{\theta}_{2im}$, $\tilde{\theta}_{3jm}$ as the maximum value of $\tilde{\theta}_{1i}$, $\tilde{\theta}_{2i}$, $\tilde{\theta}_{3j}$ respectively. A constant ϱ is chosen as

$$\varrho = \nu_{\max} + \Delta_1 \tag{7.49}$$

where Δ_1 is a positive constant to be selected and positive constant $\nu_{\max}$ is defined as follows

$$\nu_{\max} = \frac{1}{2}\tilde{\theta}_{1,m}^T \Gamma_1^{-1}\tilde{\theta}_{1,m} + \frac{1}{2}\tilde{\theta}_{2,m}^T \Gamma_2^{-1}\tilde{\theta}_{2,m} + \frac{1}{2}\tilde{\theta}_{3,m}^T \Gamma_3^{-1}\tilde{\theta}_{3,m} \tag{7.50}$$

where $\tilde{\theta}_{1,m} = [\tilde{\theta}_{11m}, ..., \tilde{\theta}_{19m}]$, $\tilde{\theta}_{2,m} = [\tilde{\theta}_{21m}, ..., \tilde{\theta}_{29m}]$ and $\tilde{\theta}_{3,m} = [\tilde{\theta}_{31m}, ..., \tilde{\theta}_{34m}]$.

Fact 7.1 τ_{us} and τ_{rs} in (7.30) are bounded and smooth in compact set Ξ, and their bounds are computable.

Proof: See Appendix A. □

Let

$$S_{us} = \max_{\mathcal{E} \in \Xi} |\tau_{us}(u, v, r)| + \Delta_2, \quad S_{rs} = \max_{\mathcal{E} \in \Xi} |\tau_{rs}(u, v, r)| + \Delta_2 \tag{7.51}$$

where Δ_2 is a positive constant to be chosen. The final output feedback controllers are designed as

$$\tau_u = \text{Sat}_{S_{us}}(\tau_{us}(\hat{u}, \hat{v}, \hat{r})), \quad \tau_v = \text{Sat}_{S_{rs}}(\tau_{vs}(\hat{u}, \hat{v}, \hat{r})) \tag{7.52}$$

where the unknown u, v and r are replaced by $\hat{u}$, $\hat{v}$ and $\hat{r}$ respectively, and the saturation function is defined as

$$\text{Sat}_{S_\star}(\tau_\star) = \begin{cases} \text{sign}(\tau_\star)S_\star & |\tau_\star| \geq S_\star \\ \tau_\star & |\tau_\star| < S_\star \end{cases}$$

where $\tau_\star$ signifies τ_{us} or τ_{rs}, $S_\star$ denotes S_{us} or S_{rs} respectively.

Now we consider the design of estimators only involving output and observer signals. For the same reasons the output feedback estimators will also be saturated. We use $\dot{\hat{\theta}}_{11}$ as an example for illustration. Let

$$S_{\hat{\theta}_{11}} = \max_{\mathcal{E} \in \Xi} |\tilde{u}\Omega_{11}| + \Delta_2. \tag{7.53}$$

Define

$$a_{\hat{\theta}_{11}} = \mathrm{Sat}_{S_{\hat{\theta}_{11}}}\left(\hat{\tilde{u}}\hat{\Omega}_{11}\right)$$

where $\hat{\tilde{u}}$ and $\hat{\Omega}_{11}$ are the values of $\tilde{u}$ and Ω_{11} obtained after the unknown u, v and r are replaced by $\hat{u}$, $\hat{v}$ and $\hat{r}$, respectively. The output feedback estimator for $\hat{\theta}_{11}$ can be designed as

$$\dot{\hat{\theta}}_{1j} = \mathrm{Proj}(a_{\hat{\theta}_{11}}, \hat{\theta}_{11}) = \begin{cases} a_{\hat{\theta}_{11}} & \text{if } \mu(\hat{\theta}_{11}) \le 0 \\ a_{\hat{\theta}_{11}} & \text{if } \mu(\hat{\theta}_{11}) \ge 0 \text{ and } \mu'(\hat{\theta}_{11}) \le 0 \\ (1 - \mu(\hat{\theta}_{11}))a_{\hat{\theta}_{11}} & \text{if } \mu(\hat{\theta}_{11}) > 0 \text{ and } \mu'(\hat{\theta}_{11}) > 0. \end{cases} \tag{7.54}$$

Other estimators can be designed in a similar way.

Notation: In the remaining part of the chapter, we use $\hat{\tilde{u}}$, $\hat{\tilde{r}}$, $\hat{\Omega}_1$, $\hat{\Omega}_2$ to denote the value of $\tilde{u}$, $\tilde{r}$, Ω_1 and Ω_2 after their unknown states u, v and r are replaced by $\hat{u}$, $\hat{v}$ and $\hat{r}$ respectively.

7.4.3 System Analysis

In this section we will establish the stability, the transient and steady-state tracking performances of the closed-loop system. The analysis can be divided into the following steps. Firstly we will prove that the velocities u, v and r and the state estimation errors are bounded with the saturated output-feedback controllers. Secondly we will show that there exists a time T_s, which can be adjusted as small as possible by choosing ε_1, such that at $t = T_s$, $\mathcal{E} \in \Xi$ and after T_s, the saturations for controllers and estimators do not take effect. Thirdly, we will prove that Ξ is an invariant set, i.e., $\mathcal{E} \in \Xi \ \forall t$, which means that the position error and orientation error are bounded. Finally we will prove that the position error and orientation error are of $O(\varepsilon)$, and the position error and the orientation error also decrease at a pre-specified exponential rate.

Now we define the observation errors as follows

$$\varphi_1 = \frac{1}{\varepsilon_1}(\phi - \hat{\phi}), \quad \varphi_2 = r - \hat{r}$$

$$\omega_1 = \frac{1}{\varepsilon_1}(\rho_2 - \hat{\rho}_2), \quad \omega_2 = u - \hat{u}$$

$$\psi_1 = \frac{1}{\varepsilon_1}(\rho_1 - \hat{\rho}_1), \quad \psi_2 = v - \hat{v} \tag{7.55}$$

and denote $\varphi = [\varphi_1 \ \varphi_2]^T$, $\omega = [\omega_1 \ \omega_2]^T$, $\psi = [\psi_1 \ \psi_2]^T$. From (7.43), (7.44), (7.46) and (7.47) we get

$$\varepsilon_1 \dot{\varphi} = \bar{A}\varphi + \varepsilon_1 \begin{bmatrix} 0 \\ G_r \end{bmatrix} \tag{7.56}$$

$$\varepsilon_1 \dot{\omega} = \bar{A}\omega + \varepsilon_1 \begin{bmatrix} 0 \\ G_u \end{bmatrix} + \begin{bmatrix} \rho_1 \varphi_2 \\ 0 \end{bmatrix} \tag{7.57}$$

$$\varepsilon_1 \dot{\psi} = \bar{A}\psi + \varepsilon_1 \begin{bmatrix} 0 \\ G_v \end{bmatrix} - \begin{bmatrix} \rho_2 \varphi_2 \\ 0 \end{bmatrix} \tag{7.58}$$

where $G_r = \dfrac{m_{11} - m_{22}}{m_{33}} uv - \dfrac{d_{33}}{m_{33}} r - \dfrac{d_{r2}}{m_{33}} |r|r - \dfrac{d_{r3}}{m_{33}} r^3 + \dfrac{1}{m_{33}} \left(\tau_r + \tau_{wr}(t) \right)$,

$G_u = \dfrac{m_{22}}{m_{11}} vr - \dfrac{d_{11}}{m_{11}} u - \dfrac{d_{u2}}{m_{11}} |u|u - \dfrac{d_{u3}}{m_{11}} u^3 + \dfrac{1}{m_{11}} \left(\tau_u + \tau_{wu}(t) \right)$, $G_v =$

$-\dfrac{m_{11}}{m_{22}} ur - \dfrac{d_{22}}{m_{22}} v - \dfrac{d_{v2}}{m_{22}} |v|v - \dfrac{d_{v3}}{m_{22}} v^3 + \dfrac{1}{m_{22}} \tau_{wv}(t)$.

Lemma 7.4. *When the systems (7.1) and (7.2) are controlled by the adaptive controllers in (7.52) with estimators in the form of (7.54), u, v, r and φ are globally bounded.*

Proof: By considering a Lyapunov function $V_\varpi = \dfrac{1}{2} \varpi^T \varpi$ where $\varpi = [\sqrt{m_{11}} u, \sqrt{m_{22}} v, \sqrt{m_{33}} r]^T$, we get

$$\dot{V}_\varpi \leq -2\xi V_\varpi + \frac{1}{4\varsigma}(\tau_u^{\ 2} + \tau_{wu\,max}^2) + \frac{1}{4\varsigma}(\tau_r^{\ 2} + \tau_{wr\,max}^2) + \frac{1}{4\varsigma}\tau_{wv\,max}^2$$

where $\xi = \min\{\dfrac{d_{11} - \varsigma}{m_{11}}, \dfrac{d_{22} - \varsigma}{m_{22}}, \dfrac{d_{33} - \varsigma}{m_{33}}\}$ and ς is a constant chosen such that $\xi > 0$. So u, v and r are globally bounded since the torques are globally saturated. Also as τ_u and τ_r are bounded, there exists positive constants g_u, g_v and g_r so that

$$|G_u| < g_u, \quad |G_v| < g_v, \quad |G_r| < g_r.$$

Since $\bar{A}$ is a Hurwitz matrix, so (7.56) is a stable system and φ is bounded. $\square$

Lemma 7.5. *For any ε_1, there exists a finite time $T_1(\varepsilon_1)$, so that for $t > T_1$, we have*

$$\| \varphi \| < c_1 \varepsilon_1 \tag{7.59}$$

where c_1 is a positive constant that is computable.

Proof: The proof is similar to that of Lemma 6.6.

Case 1: For $\nu(t) \geq \dfrac{4\iota^2}{\gamma^2}\varepsilon_1^2$ we obtain

$$\dot{\nu} \leq -\frac{\gamma}{2\varepsilon_1}\nu \tag{7.60}$$

which gives that $\nu(t) \leq \nu(0)e^{-\frac{\gamma}{2\varepsilon_1}t}$. For any bounded initial conditions of ϕ, $\hat{\phi}$, r and $\hat{r}$, from the definition of φ in (7.55) we know $\nu(0)$ can be written as $\nu(0) = \dfrac{l_3}{\varepsilon_1^2}$ where $l_3 = p_{11}s^2 + (p_{21} + p_{12})\varepsilon_1 sq + p_{22}q^2\varepsilon_1^2$, $s = \phi(0) - \hat{\phi}(0)$, $q = r(0) - \hat{r}(0)$. Thus we get $\nu(t) \leq \dfrac{l_3}{\varepsilon_1^2}e^{-\frac{\gamma}{2\varepsilon_1}t}$.

So when the initial condition $\dfrac{l_3}{\varepsilon_1^2}$ satisfies that $\dfrac{l_3}{\varepsilon_1^2} > \dfrac{4\iota^2}{\gamma^2}\varepsilon_1^2$, the time $T_1(\varepsilon_1)$ for ν to reach the ball $\nu \leq \dfrac{4\iota^2}{\gamma^2}\varepsilon_1^2$ from $\nu(0)$ is

$$T_1(\varepsilon_1) = \frac{2\varepsilon_1}{\gamma}\left(\ln\frac{l_4}{\varepsilon_1^4}\right) \tag{7.61}$$

where $l_4 = \dfrac{l_3\gamma^2}{4\iota^2}$. Since (7.61) tends to zero as $\varepsilon_1 \to 0$, so we can choose a sufficiently small $T_1(\varepsilon_1)$ such that $\forall t > T_1(\varepsilon_1)$; we have $\nu \leq \dfrac{4\iota^2}{\gamma^2}\varepsilon_1^2$, which means $\|\varphi\| \leq c_1\varepsilon_1$, where $c_1 = \sqrt{\dfrac{4\iota^2}{\gamma^2\lambda_{\min}(P)}}$.

Case 2: When $\nu(t) \leq \dfrac{4\iota^2}{\gamma^2}\varepsilon_1^2$, the conclusion holds automatically for any $T_1(\varepsilon_1) \geq 0$. This completes the proof.   □

Fact 7.2 *The upper-bound of $\dot{V}_2$ within Ξ along (7.52) and (7.54) is computable.*

Proof: See Appendix.   □

Lemma 7.6. *There exists a T_s, which can be adjusted arbitrarily small, such that for $t \in [T_s, T_e]$ the saturation for controllers (7.52) and estimators (7.54) will not take effect, where T_e is the first time instant that $\mathcal{E}$ leaves the set Ξ.*

Remark 7.7. In Lemma 7.8 given later we will show that $T_e = \infty$ and thus Ξ is an invariant set.

Proof: Based on Fact 7.2, we can find a constant c_4 such that $|\dot{V}_2| \leq c_4$ before $\mathcal{E}$ exits Ξ. So we have

$$V_2(T_1) \le V_2(0) + \int_0^{T_1(\varepsilon_1)} c_4 dt = V_2(0) + c_4 T_1(\varepsilon_1).$$

From Lemma 7.5, if we choose an ε_1 such that

$$V_2(0) + c_4 T_1(\varepsilon_1) < \varrho \tag{7.62}$$

then for $t \in [0, T_1(\varepsilon_1)]$, we have $\mathcal{E} \in \Xi$.

Suppose T_e is the first time that $\mathcal{E}$ exits from Ξ. So for $t \in (0, T_e]$, we get $|\chi| < \sqrt{2\varrho}$ and $|\gamma| < \sqrt{2\varrho}$, which means $|z_e| < \eta_0 S_1(\chi)$ and $|\phi_e| < \eta_0 S_2(\gamma)$. Thus

$$|x| < |x_d| + |x_e| < a^\star + |z_e|$$
$$|y| < |y_d| + |y_e| < a^\star + |z_e|.$$

Finally we get

$$|\rho_1| < |x| + |y| < 2a^\star + 2|z_e|, \quad |\rho_2| < |x| + |y| < 2a^\star + 2|z_e|. \tag{7.63}$$

From (7.46) and (7.47), ω and ψ are bounded for $t \in (0, T_e]$. Following the similar proofs of Lemma 7.4 and Lemma 7.5, there exist $T_2(\varepsilon_1)$ and $T_3(\varepsilon_1)$, which can be arbitrarily small if ε_1 is small enough, such that for $T_e > t > T_2(\varepsilon_1)$ and $T_e > t > T_3(\varepsilon_1)$, we get

$$\| \omega \| < c_2 \varepsilon_1, \quad \| \psi \| < c_3 \varepsilon_1 \tag{7.64}$$

respectively, where c_2 and c_3 are positive constants.

Fact 7.3 *There are two positive constants c_u and c_r such that*

$$\tau_{us}(\hat{u}, \hat{v}, \hat{r}) < \tau_{us}(u, v, r) + c_u \varepsilon_1, \quad \tau_{rs}(\hat{u}, \hat{v}, \hat{r}) < \tau_{rs}(u, v, r) + c_r \varepsilon_1. \tag{7.65}$$

when $t \in [T_s, T_e]$, where

$$T_s = \max\{T_1(\varepsilon_1), T_2(\varepsilon_1), T_3(\varepsilon_1)\} \tag{7.66}$$

Proof: See Appendix C. □

Since $\mathcal{E} \in \Xi$ for $t \in (0, T_e]$, we have $\tau_{us}(u, v, r) < \max_{\mathcal{E} \in \Xi} |\tau_{us}(u, v, r)|$ and $\tau_{rs}(u, v, r) < \max_{\mathcal{E} \in \Xi} |\tau_{rs}(u, v, r)|$. Now choose ε_1 such that

$$c_u \varepsilon_1 < \Delta_2, \quad c_r \varepsilon_1 < \Delta_2. \tag{7.67}$$

Then from (7.51) we have

$$\tau_{us}(\hat{u}, \hat{v}, \hat{r}) < S_{us}, \quad \tau_{rs}(\hat{u}, \hat{v}, \hat{r}) < S_{rs}$$

which means that the saturation does not take effect when $t \in [T_s, T_e]$. With the same analysis, for parameter estimators, the saturation will not take effect for $t \in [T_s, T_e]$ either. The following fact is given.

Fact 7.4 *The time derivative of V_2 in (7.28) along (7.52) and (7.54) satisfies*

$$\dot{V}_2 \leq -k_1\chi^2 - k_2\gamma^2 - (k_3 + d_{11})\tilde{u}^2 - (k_4 + d_{33})\tilde{r}^2 + 0.2785\sum_{i=1}^{4}\epsilon_i + k\varepsilon_1$$

(7.68)

for $t \in [T_s, T_e]$ where k is a known positive constant depending on the initial condition and Δ_1.

Proof: See Appendix D.   □

Next we will show that T_e in Lemma 7.6 is ∞, namely $\mathcal{E}$ will never exit Ξ and thus Ξ is an invariant set.

Lemma 7.8. *Ξ is an invariant set.*

Proof: Define a new set

$$\Psi = \left\{e : k_1\chi^2 + k_2\gamma^2 + (k_3 + d_{11})\tilde{u}^2 + (k_4 + d_{33})\tilde{r}^2 \leq 0.2785\sum_{i=1}^{4}\epsilon_i + k\varepsilon_1\right\}$$

(7.69)

where e is defined in (7.40). Also denote κ as

$$\kappa = \max_{e \in \Psi} V_e$$

(7.70)

where $V_e = V_1 + \dfrac{m_{11}}{2}\tilde{u}^2 + \dfrac{m_{33}}{2}\tilde{r}^2 = V_2 - \dfrac{1}{2}\sum_{i=1}^{3}\tilde{\theta}_i^T \Gamma_i^{-1}\tilde{\theta}_i$. From (7.28) and (7.69), the design parameter ε_1 and ϵ_i, $i = 1, ..., 4$ can be chosen to make κ satisfy

$$\kappa < \frac{1}{2}\Delta_1.$$

(7.71)

From the definitions of V_e and Ψ we know this can be easily done. We now divide Ξ into two set Ξ_1 and Ξ_2 as follows

$$\Xi_1 = \{\mathcal{E} : e \in \Psi;\ \mathcal{E} \in \Xi\} \quad \Xi_2 = \{\mathcal{E} : e \notin \Psi;\ \mathcal{E} \in \Xi\}.$$

Clearly $\Xi = \Xi_1 \cup \Xi_2$ and $\Xi_1 \cap \Xi_2 = \emptyset$.

Ξ is divided so that when $\mathcal{E} \in \Xi_1$, $V_2 < V_e + \nu_{max} < \dfrac{1}{2}\Delta_1 + \nu_{max} = \varrho - \dfrac{1}{2}\Delta_1$, and when $\mathcal{E} \in \Xi_2$, $\dot{V}_2 < 0$ from (7.68) and (7.69).

When $\mathcal{E} \in \Xi_1$, $\dot{V}_2$ may be positive. But since $\dot{V}_2 < c_4$, $V_2(t)$ is continuous. Therefore if $\mathcal{E}$ exits Ξ_1, it will only enter Ξ_2, because $V_2(t) < \varrho - \dfrac{1}{2}\Delta_1$ when $\mathcal{E} \in \Xi_1$. Thus when $\varrho - \dfrac{1}{2}\Delta_1 \leq V_2(t) < \varrho$, $\mathcal{E}$ is in Ξ_2. In other words, $\mathcal{E}$ will not exit Ξ from Ξ_1 directly.

Note that T_e defined in Lemma 7.6 is the first time that $\mathcal{E}$ exits Ξ satisfying $V_2(T_e) = \varrho$. Then $V_2(T_e - \Delta t) < \varrho$, where $0 < \Delta t < T_e$ is an arbitrarily small constant. As stated, $\mathcal{E}(T_e - \Delta t) \in \Xi_2$, so $\dot{V}_2(T_e - \Delta t) < 0$ and thus as Δt decreases $V_2(T_e - \Delta t)$ is monotonically decreasing. This contradicts that $V_2(T_e) = \varrho$. So $\mathcal{E}$ will not exit Ξ from Ξ_2 either, i.e. $T_e = \infty$. Therefore Ξ is an invariant set. This completes the proof. $\square$

Since Ξ is an invariant set, then from Lemma 7.5 and Lemma 7.6,

$$\tau_{us}(\hat{u}, \hat{v}, \hat{r}) < S_{us}, \; \tau_{rs}(\hat{u}, \hat{v}, \hat{r}) < S_{rs}$$

for $t \in [T_s, \infty)$, which means the saturation will not take effect after T_s, and we have

$$\dot{V}_2 \leq -k_1 \chi^2 - k_2 \gamma^2 - (k_3 + d_{11})\tilde{u}^2 - (k_4 + d_{33})\tilde{r}^2 + \varepsilon \tag{7.72}$$

where $\varepsilon = 0.2785 \sum_{i=1}^{4} \epsilon_i + k\varepsilon_1$ for $t \in [T_s, \infty)$. Now based on Lemmas 7.4-7.8, we have the following theorem.

Theorem 7.9. *Consider the system consisting of the underactuated ship system (7.1) and (7.2), the adaptive output feedback controllers (7.52) with observers (7.44), (7.46) and (7.47), and parameter estimators (7.54). Under Assumptions 7.1, 7.2 and 7.3, and assuming $\mathcal{E}(0) \in \Xi$, then there exists a computable ε_1^* satisfying (7.60), (7.64) and (7.67), such that for all $\varepsilon_1 \in (0, \varepsilon_1^*)$, (i) the signals in the closed-loop systems are bounded; (ii) the mean-square tracking errors are of $O(\varepsilon)$. (iii) the position error z_e and orientation error ϕ_e converge to arbitrarily small residual sets at a pre-specified exponential rate.*

Proof: (i) Since Ξ is an invariant set, it is obvious that all signals contained in the closed-loop system are bounded.

(ii) Due to the parameter projection, $\tilde{\theta}_{1i}$, $\tilde{\theta}_{2i}$ and $\tilde{\theta}_{3j}$, $1 \leq i \leq 9$, $1 \leq j \leq 4$ are bounded. From (7.72) we know

$$k_{\min}\left(\chi^2 + \gamma^2 + \tilde{u}^2 + \tilde{r}^2\right) + \dot{V}_2(t) \leq \varepsilon \tag{7.73}$$

where $k_{\min} = \min(k_1, \; k_2, k_3 + d_{11}, k_4 + d_{33})$. Integrating (7.73), and using the fact the $V_2(t)$ is bounded, we get

$$\lim_{T \to \infty} \frac{1}{T} \int_0^T e^T e \leq \frac{\varepsilon}{k_{\min}}$$

which means that the mean-square position errors and virtual control errors are of the order of $O(\varepsilon)$, i.e., e can be adjusted arbitrarily small if ε is small enough. From (7.15), the properties of $S_1(\chi), S_2(\gamma)$ and (7.11), we obtain that

$$\sigma \le \alpha_1\eta_\infty < z_e(t) < \beta_1\eta_0 \le \delta$$

$$-\frac{\pi}{2} + \sigma_2 \le -\alpha_2\eta_0 < \phi_e(t) < \beta_2\eta_0 \le \frac{\pi}{2} - \sigma_2, \quad \forall\ t \ge 0. \tag{7.74}$$

if $z_e(0) \in \left(\alpha_1\eta_\infty, \beta_1\eta_0\right)$ and $\phi_e(0) \in (-\pi/2+\sigma_2, \pi/2-\sigma_2)$, which means (7.7) can always be satisfied.

We also have $\lim_{t\to\infty} \phi_e(t) = \varepsilon_\star$ and $\lim_{t\to\infty} z_e(t) = (\alpha_1 + \epsilon + \varepsilon^\star)\eta_\infty \in \left(\alpha_1\eta_\infty, \beta_1\eta_\infty\right)$ where $\varepsilon_\star$ and $\varepsilon^\star$ are unknown constants that can be arbitrarily small if ε is small enough, due to the continuities of $S_1(\chi)$ and $S_2(\gamma)$ and the fact that χ and γ are arbitrarily small. It implies that the underactuated ship can follow its reference trajectory with arbitrarily small position error z_e and orientation error ϕ_e.

(iii) From (7.9)

$$z_e < \beta_1\left((\eta_0 - \eta_\infty)e^{-at} + \eta_\infty\right), \phi_e < \beta_2\left((\eta_0 - \eta_\infty)e^{-at} + \eta_\infty\right) \tag{7.75}$$

which means the position error z_e and orientation error ϕ_e will converge at a pre-specified exponential speed governed by positive constant a. This completes the proof. $\qquad\square$

Remark 7.10. The keys for solving the adaptive output feedback control of the underactuated ships lie in several aspects. Firstly observers for v, v and r are designed separately to avoid the nonholonomic constraint. Secondly saturation is introduced to the output feedback controller and estimators. This will prevent the peaking phenomenon from entering the robot system. Thirdly a compact set Ξ of $\mathcal{E}$ is set up. The stability of the whole system is guaranteed by proving Ξ is an invariant set.

Remark 7.11. Our methodology is not only limited to the underactuated ships, but also applicable to the unsolved adaptive output feedback control problem of many other uncertain mechanical systems with nonholonomic constraints, such as underwater vehicles, VTOL aircrafts, aerospace vehicle etc.

7.5 Simulation Results

In this section, we present our simulation studies with a monohull ship having one propeller and one rudder. The parameters of the ship are given in [21] as follows: $m_{11} = 120 \times 10^3\text{kg}$, $m_{22} = 177.9 \times 10^3\text{kg}$, $m_{33} = 636 \times 10^5\text{kg}$, $d_{11} = 215 \times 10^2\text{kg}$, $d_{22} = 147 \times 10^3\text{kg}$, $d_{33} = 802 \times 10^4\text{kg}$, $d_{u2} = 0.2d_{11}$, $d_{u3} = 0.1d_{11}$, $d_{v2} = 0.2d_{22}$, $d_{v3} = 0.1d_{22}$, $d_{r2} = 0.2d_{33}$, $d_{r3} = 0.1d_{33}$. These parameters are assumed to be unknown. Their maximum and minimum values are set to fluctuate 30% above and below their true values. The bounds of the external disturbances are assumed to be $\tau_{wu} = 10 \times 10^3(1 + \text{rand}(\cdot))$, $\tau_{wv} = 10 \times 10^3(1 + \text{rand}(\cdot))$, $\tau_{wr} = 10 \times 10^4(1 + \text{rand}(\cdot))$. where $\text{rand}(\cdot)$ is random noise with magnitude bounded by 1 and zero from below. This choice results in nonzero-mean disturbances.

The design parameters are chosen as follows: $k_1 = 2$, $k_2 = 3$, $k_3 = 8$, $k_4 = 8$, $\epsilon_1 = 0.2$, $\epsilon_2 = 0.2$, $\epsilon_3 = 0.1$, $\epsilon_4 = 4$, $\epsilon_* = 0.2$, $\epsilon^* = 5$, $u_0 = 10$, $\varepsilon_1 = 0.05$, $\eta_0 = 2$, $\eta_\infty = 0.1$, $a = 2$, $\alpha_1 = 1$, $\beta_1 = 40$, $\alpha_2 = 0.7$, $\beta_2 = 0.7$, $\Gamma_i^{-1} = 10I$, for $i = 1, 2, 3$. The reference trajectories are taken as $\{x_{od}, y_{od}\} = [s, 100\sin(0.05s)]^T$. The initial conditions are set as $\{x(0), y(0), \phi(0), u(0), v(0), r(0)\} = \{-35, -10, 1.2, 0, 0, 0\}$. All initial values of parameter estimates are taken to be 60% of their assumed true values. The saturation bound calculated is 1.2×10^7N. The results on the ship position, the tracking errors and the the torques for τ_u and τ_r evolving with time are shown in Fig.7.2, Fig.7.3, Fig.7.4, Fig.7.5 and Fig.7.6, respectively. It can be seen from the figures that the references can be followed with arbitrarily small position error and orientation error. Fig.7.7 and Fig.7.8 show the estimates of $\hat{\theta}_1$ and $\hat{\theta}_3$. Fig.7.9 - Fig.7.11 show the estimate errors of u, v and r.

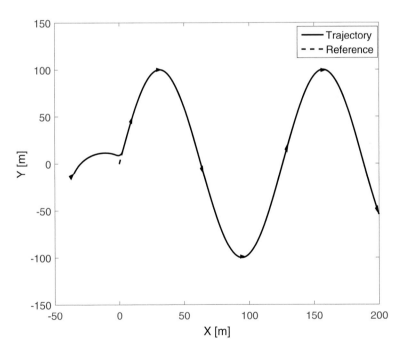

Fig. 7.2. Position and orientation of ship following a path $y = 100\sin(0.05x)$

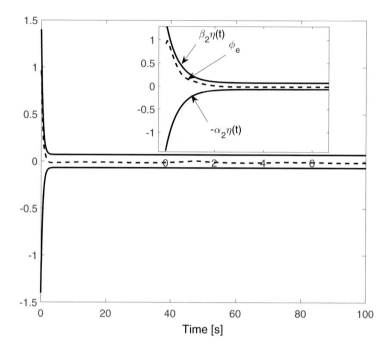

Fig. 7.3. Orientation error $\phi_e(t)$.

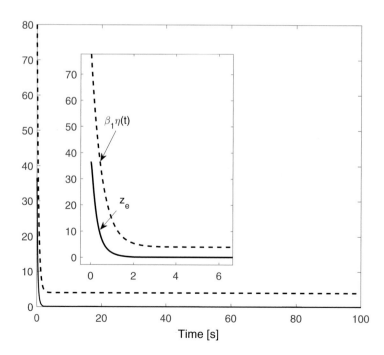

Fig. 7.4. Position error $z_e(t)$.

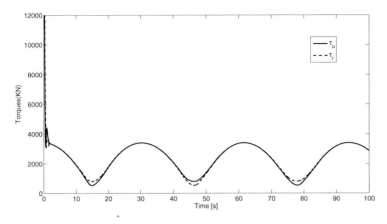

Fig. 7.5. Torques τ_u and τ_r with saturation limit 1.2×10^7 N.

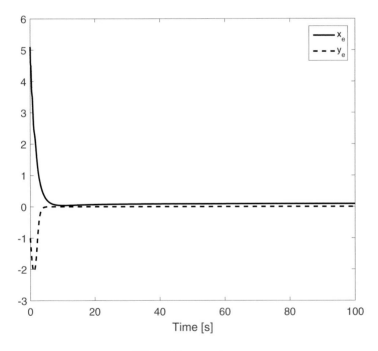

Fig. 7.6. x_e and y_e

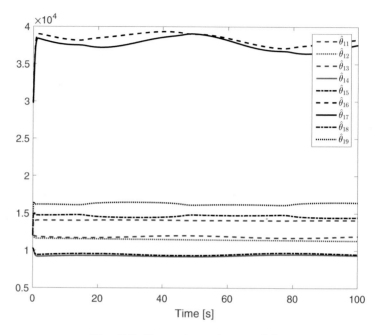

Fig. 7.7. Parameter estimator of θ_1.

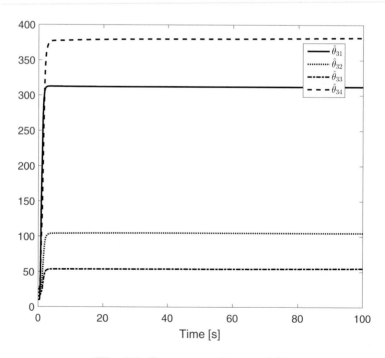

Fig. 7.8. Parameter estimator of θ_3.

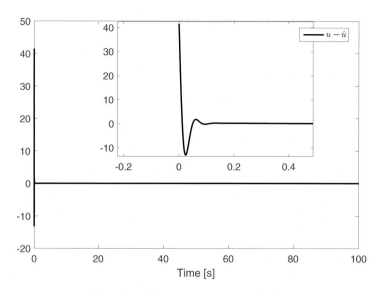

Fig. 7.9. Observer estimator of u-$\hat{u}$.

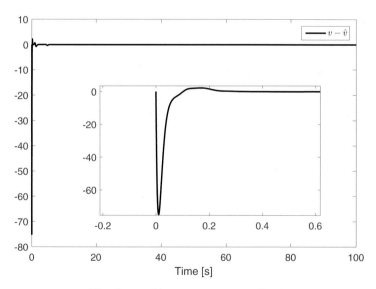

Fig. 7.10. Observer estimator of v-$\hat{v}$

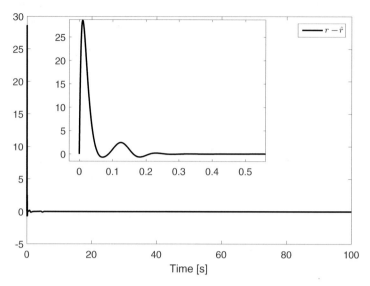

Fig. 7.11. Observer estimator of r-$\hat{r}$.

7.6 Conclusions

In this chapter, path following control of underactuated ships with unknown system parameters and external disturbances induced by wave, wind and ocean current is considered. By using the prescribed performance bound technique, we develop both state-feedback and output-feedback robust adaptive controllers. Besides showing system stability, the mean-square tracking errors are of order of an arbitrarily small design parameter, and the position error and the orientation error converge to arbitrarily small residual sets at a pre-specified exponential rate.

7.7 Appendix

7.7.1 Proof of Fact 7.1

From (7.30) τ_{us} and τ_{rs} are composed by $\tilde{u}$, $\tilde{r}$, ξ_1, ξ_2, χ, γ, ϕ_e, z_e, $\hat{\theta}_{1i}$, $\hat{\theta}_{2i}$, $\hat{\theta}_{3j}$, $i = 1, ..., 9$, $j = 1, ..., 4$, $\dot{s}$, η, $\dot{\eta}$, x'_d, y'_d, $\sin(\phi)$ and $\cos(\phi)$. From Assumption 7.1 and parameter projection, we know all these elements are bounded with computable upper-bounds (or within Ξ). So from (7.30) the upper-bounds of τ_{us} and τ_{rs} are computable within Ξ, i.e., there are two positive constants S'_{us} and S'_{rs} such that

$$\tau_{us}(u, v, r)|_{\mathcal{E} \in \Xi} \leq S'_{us}, \quad \tau_{rs}(u, v, r)|_{\mathcal{E} \in \Xi} \leq S'_{rs}$$

7.7.2 Proof of Fact 7.2

From (7.68)

$$\dot{V}_2 = -k_1\chi^2 - k_2\gamma^2 - \chi\xi_1\cos(\phi_e)\tilde{u} + \xi_2\gamma\tilde{r} - \xi_2\gamma\Big(\cos(\phi)f_1 - \sin(\phi)f_2\Big)\tilde{u}$$

$$+ m_{11}\tilde{u}\dot{\tilde{u}} + m_{33}\tilde{r}\dot{\tilde{r}} - \sum_{i=1}^{3}\tilde{\theta}_i^T\Gamma_i^{-1}\dot{\tilde{\theta}}_i. \tag{7.76}$$

Since τ_u, τ_r and $\dot{\hat{\theta}}_i$ are globally saturated. Thus from (7.27) and the proof of Fact 7.1 the upper-bound of $\dot{V}_2$ is computable; i.e. there exists a positive constant c_4 such that

$$\dot{V}_2 < c_4.$$

7.7.3 Proof of Fact 7.3

From (7.55), $|u - \hat{u}| < c_2\varepsilon_1$, $|v - \hat{v}| < c_3\varepsilon_1$, $|r - \hat{r}| < c_1\varepsilon_1$ when $\mathcal{E} \in \Xi$. Thus from (7.37), $\hat{u}_d = a_1\hat{v} + a_2 < u_d + a_1c_3\varepsilon_1$, and similarly $\hat{r}_d < r_d + a_3c_3\varepsilon_1$. Taking the first item of Ω_1 as an example, $\hat{v}\hat{r} - vr = \hat{v}\hat{r} - \hat{v}r + \hat{v}r - vr < \Big(|\hat{v}|c_1 + |r|c_3\Big)\varepsilon_1 < \Big(|v|c_1 + |r|c_3 + c_1c_3\varepsilon_1\Big)\varepsilon_1 := \bar{c}\varepsilon_1$ where $\bar{c}$ is a known positive constant due to the fact that the upper-bounds of v and r are known. It can be similarly checked that this conclusion holds for each item in Ω_1 and Ω_2 when $\mathcal{E} \in \Xi$.

Suppose $\tilde{u} - \hat{\tilde{u}} < c_w\varepsilon_1$ where c_w is a positive constant, then based on the mean value theorem,

$$\tanh\Big(\frac{\tilde{u}\hat{\theta}_{31}}{\epsilon_1}\Big) - \tanh\Big(\frac{\hat{\tilde{u}}\hat{\theta}_{31}}{\epsilon_1}\Big) = \mathrm{sech}^2\Big(\frac{\tilde{u}_x\hat{\theta}_{31}}{\epsilon_1}\Big)\frac{\hat{\theta}_{31}}{\epsilon_1}(\tilde{u} - \hat{\tilde{u}}) \leq \frac{\hat{\theta}_{31}c_w}{\epsilon_1}\varepsilon_1$$

where $\tilde{u}_x \in (\tilde{u}, \hat{\tilde{u}})$. Similar conclusions also hold for other $\tanh(\cdot)$ in τ_{us} and $\tau_r s$ when $\mathcal{E} \in \Xi$. Thus from (7.30) we can make the conclusion that when $\mathcal{E} \in \Xi$ there are two positive constants c_u and c_r such that

$$|\tau_{us}(\hat{u}, \hat{v}, \hat{r}) - \tau_{us}(u, v, r)| < c_u\varepsilon_1, \quad |\tau_{rs}(\hat{u}, \hat{v}, \hat{r}) - \tau_{rs}(u, v, r)| < c_r\varepsilon_1. \tag{7.77}$$

when $t \in [T_s, T_e]$. Thus (7.65) holds.

7.7.4 Proof of Fact 7.4

From (7.77), $\tilde{u}\tau_{us}(\hat{u}, \hat{v}, \hat{r}) < \tilde{u}\tau_{us}(u, v, r) + |\tilde{u}|c_u\varepsilon_1$ and $\tilde{r}\tau_{rs}(\hat{u}, \hat{v}, \hat{r}) < \tilde{r}\tau_{rs}(u, v, r) + |\tilde{r}|c_r\varepsilon_1$. Thus from (7.27) and (7.28)

$$\dot{V}_2 = -k_1\chi^2 - k_2\gamma^2 - \chi\xi_1\cos(\phi_e)\tilde{u} + \xi_2\gamma\tilde{r} - \xi_2\gamma\Big(\cos(\phi)f_1 - \sin(\phi)f_2\Big)\tilde{u}$$

$$+ m_{11}\tilde{u}\dot{\tilde{u}} + m_{33}\tilde{r}\dot{\tilde{r}} - \sum_{i=1}^{3}\tilde{\theta}_i^T \Gamma_i^{-1}\dot{\hat{\theta}}_i$$

$$< -k_1\chi^2 - k_2\gamma^2 - (k_3 + d_{11})\tilde{u}^2 - (k_4 + d_{33})\tilde{r}^2 + |\tilde{u}|c_u\varepsilon_1 + |\tilde{r}|c_r\varepsilon_1 + \tilde{u}\tilde{\theta}_1^T\Omega_1$$

$$+ \tilde{r}\tilde{\theta}_2^T\Omega_2 + \tilde{u}\tau_{wu}(t) + \tilde{u}\frac{m_{11}}{m_{22}}\frac{\partial u_d}{\partial v}\tau_{wv}(t) + \tilde{r}\tau_{wr}(t)$$

$$+ \tilde{r}\frac{m_{33}}{m_{22}}\frac{\partial r_d}{\partial v}\tau_{wv}(t) - \sum_{i=1}^{3}\tilde{\theta}_i^T \Gamma_i^{-1}\dot{\hat{\theta}}_i$$

$$- \tilde{u}\hat{\theta}_{31}\tanh(\frac{\tilde{u}\hat{\theta}_{31}}{\epsilon_1}) - \tilde{u}\hat{\theta}_{32}\frac{\partial u_d}{\partial v}\tanh(\frac{\partial u_d}{\partial v}\frac{\tilde{u}\hat{\theta}_{32}}{\epsilon_2}) - \tilde{r}\hat{\theta}_{33}\tanh(\frac{\tilde{r}\hat{\theta}_{33}}{\epsilon_3})$$

$$- \tilde{r}\hat{\theta}_{34}\frac{\partial r_d}{\partial v}\tanh(\frac{\partial r_d}{\partial v}\frac{\tilde{r}\hat{\theta}_{34}}{\epsilon_4}).$$

Obviously it has

$$\tilde{u}\tau_{wu}(t) + \tilde{u}\frac{m_{11}}{m_{22}}\frac{\partial u_d}{\partial v}\tau_{wv}(t) + \tilde{r}\tau_{wr}(t) + \tilde{r}\frac{m_{33}}{m_{22}}\frac{\partial r_d}{\partial v}\tau_{wv}(t) \le |\tilde{u}|\theta_{31} + |\tilde{u}\frac{\partial u_d}{\partial v}|\theta_{32}$$

$$+ |\tilde{r}|\theta_{33} + |\tilde{r}\frac{\partial r_d}{\partial v}|\theta_{34}.$$

From (7.33) we have

$$|\tilde{u}|\hat{\theta}_{31} + |\tilde{u}\frac{\partial u_d}{\partial v}|\hat{\theta}_{32} + |\tilde{r}|\hat{\theta}_{33} + |\tilde{r}\frac{\partial r_d}{\partial v}|\hat{\theta}_{34} - \tilde{u}\hat{\theta}_{31}\tanh(\frac{\tilde{u}\hat{\theta}_{31}}{\epsilon_1})$$

$$- \tilde{u}\hat{\theta}_{32}\frac{\partial u_d}{\partial v}\tanh(\frac{\partial u_d}{\partial v}\frac{\tilde{u}\hat{\theta}_{32}}{\epsilon_2})$$

$$- \tilde{r}\hat{\theta}_{33}\tanh(\frac{\tilde{r}\hat{\theta}_{33}}{\epsilon_3}) - \tilde{r}\hat{\theta}_{34}\frac{\partial r_d}{\partial v}\tanh(\frac{\partial r_d}{\partial v}\frac{\tilde{r}\hat{\theta}_{34}}{\epsilon_4}) \le 0.2785\sum_{i=1}^{4}\epsilon_i.$$

Thus

$$\dot{V}_2 \le -k_1\chi^2 - k_2\gamma^2 - (k_3 + d_{11})\tilde{u}^2 - (k_4 + d_{33})\tilde{r}^2 + |\tilde{u}|c_u\varepsilon_1 + |\tilde{r}|c_r\varepsilon_1 + \hat{\tilde{u}}\tilde{\theta}_1^T\Omega_1$$

$$+ \hat{\tilde{r}}\tilde{\theta}_2^T\Omega_2 + |\hat{\tilde{u}}|\tilde{\theta}_{31} + |\hat{\tilde{u}}\frac{\partial u_d}{\partial v}|\tilde{\theta}_{32} + |\hat{\tilde{r}}|\tilde{\theta}_{33} + |\hat{\tilde{r}}\frac{\partial r_d}{\partial v}|\tilde{\theta}_{34} - \sum_{i=1}^{3}\tilde{\theta}_i^T \Gamma_i^{-1}\dot{\hat{\theta}}_i + \underline{k}\varepsilon_1$$

where $\underline{k}$ is a positive constant. Due to the choices of $\dot{\hat{\theta}}_i$ in (7.54), and the fact that for $t \in [T_s, T_e]$, the saturation will not take effectiveness for $\dot{\hat{\theta}}_i$, thus for $t \in [T_s, T_e]$,

$$\dot{V}_2 \le -k_1\chi^2 - k_2\gamma^2 - (k_3 + d_{11})\tilde{u}^2 - (k_4 + d_{33})\tilde{r}^2 + 0.2785\sum_{i=1}^{4}\epsilon_i + \underline{k}\varepsilon_1.$$

Lemma 7.12. *[8] The following projection*

$$Proj\{\tau\} = \begin{cases} \tau & \hat{\theta} \in \Pi \quad or \ \nabla_{\hat{\theta}}P^T\tau \le 0 \\ (I - \Gamma\frac{\nabla_{\hat{\theta}}P\nabla_{\hat{\theta}}P^T}{\nabla_{\hat{\theta}}P^T\Gamma\nabla_{\hat{\theta}}P})\tau & \hat{\theta} \in \partial\Pi \ or \ \nabla_{\hat{\theta}}P^T\tau > 0 \end{cases} \qquad (7.78)$$

has the following properties if $|\hat{b}(t_0)| \le b_M$:

(1) $|\hat{b}(t)| \le b_M + \epsilon$, $\forall 0 \le t_0 \le t < \infty$.

(2) $Proj(a, \hat{b})$ is Lipschitz continuous.

(3) $|Proj(a, \hat{b})| \le |a|$

(4) $\tilde{b}Proj(a, \hat{b}) \ge \tilde{b}a$ where $\tilde{b} = b - \hat{b}$.

Part III

Adaptive Fault-Tolerant Control of
Underactuated Nonlinear Systems

8

Adaptive Fault-Tolerant Control of Underactuated Ships with Actuator Redundancy

In this chapter, a nonlinear adaptive fault-tolerant control strategy is developed to force an underactuated ship with thruster redundancy to follow a predefined path, despite the presence of unknown system parameters and environmental disturbances induced by wave, wind and ocean current. The techniques involved in the design and analysis include using the backstepping, parameter projection techniques and a traverse function. It is shown that with our proposed controller, the reference path can be tracked globally with an arbitrarily small tracking error. Simulation results demonstrate the effectiveness of our proposed controllers.

8.1 Introduction

The control of underactuated systems with nonholonomic constraints has received vast attention from the control community over the past few decades. Typical examples for this type of systems include nonholonomic mobile robots, underactuated ships, underwater vehicles and vertical take-off and landing (VTOL) aircrafts *etc*. Note there exists no continuous time-invariant feedback controller that can ensure the stabilization of this type of systems, due to the fact that the Brockett's necessary condition [1] for stabilization is not met. In [16], the design method developed for chained-form systems was adopted for underactuated ships through a coordinate transformation. Then exponential stability of a closed-loop system was ensured and the position variable as well as the course angle of the ship were steered to track the reference trajectory. In [73, 20], a continuous time-varying tracking controller was designed to yield globally uniformly ultimately bounded tracking by transforming the ship tracking system into a skew-symmetric form and designing a time-varying dynamic oscillator. In [42], a simple state-feedback control law was exploited to render the tracking error globally $\mathcal{K}$-exponentially stable. In [45], it was shown that exponential stabilization for an autonomous underwater vehicle (AUV) with only four actuators is possible by utilizing previous results on attitude stabilization for a spacecraft [92]. In [21], a global robust adaptive controller was proposed to force an underactuated ship to follow a reference path by considering both constant and time-varying disturbances. In [9], a globally tracking controller was developed under the case that the mass and damping matrices are not diagonal. Moreover, the reference trajectory was not necessary to be generated by a ship model. In [25], by introducing global nonlinear coordinate changes to transform the ship dynamics to that affine in the ship velocities, global partial-state feedback and output-feedback control schemes for tracking control of an underactuated surface ship were proposed. In [17], two tracking solutions were presented using Lyapunov's direct and passivity approaches for underactuated ships.

A vital problem in the control of underactuated ships is that failure occurrence on the actuators [74, 75] may cause the control failure and even lead to disasters. Such failures are often uncertain in time, value and pattern, and may cause instability or even catastrophic accidents. For the sake of reliability and safety, actuator failure compensation has received an increasing amount of attention and considerable achievements have been made based on various approaches such as robust control [76], multiple-model [77, 78], sliding mode control [79, 80, 81], and so on. The thruster-force allocation problem is important to overcome thruster failure[85]. The allocation of thruster forces of an AUV which can accommodate both thruster faults and saturation was presented in [86, 87]. A fault diagnosis and accommodation system was proposed using weighted pseudoinverse in [88]. Weighted pseudoinverse and quantum particle swarm optimization were introduced for hybrid fault-tolerant control in [89]. The combination of backstepping technique and fuzzy fault-tolerant control was presented for tracking control of fully actuated surface vessels in [90]. However, all these works focused on fully or overactuated systems. This implies that the cost and weight of the system are increased because there must be more devices. Moreover, if a fully actuated system is damaged, we then have an underactuated controller [91]. Therefore, it is necessary to develop a fault-tolerant controller for underactuated systems.

In this chapter, we shall propose an adaptive fault-tolerant control design method for the underactuated ships with a thruster in the presence of external disturbances. In contrast to the aforementioned results, all the system parameters are allowed to be unknown. By employing the traverse function approach [15], an "auxiliary manipulated variable" is introduced, with which the difficulty encountered in controlling an underactuated system can be overcome.

8.2 Problem Formulation

Similar to [16], we consider the ship as shown in Fig.8.1, of which the dynamic model can be described as follows

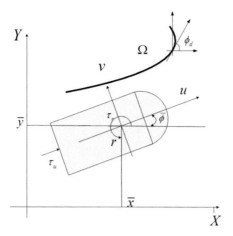

Fig. 8.1. The coordinates of an underacted ship.

$$\begin{bmatrix} \dot{\bar{x}} \\ \dot{\bar{y}} \\ \dot{\bar{\phi}} \end{bmatrix} = \begin{bmatrix} \cos\bar{\phi} & -\sin\bar{\phi} & 0 \\ \sin\bar{\phi} & \cos\bar{\phi} & 0 \\ 0 & 0 & 1 \end{bmatrix} \begin{bmatrix} u \\ v \\ r \end{bmatrix} \tag{8.1}$$

$$\dot{u} = \frac{m_{22}}{m_{11}} vr - \frac{d_{11}}{m_{11}} u - \sum_{i=2}^{3} \frac{d_{ui}}{m_{11}} |u|^{i-1} u + \frac{1}{m_{11}} \sum_{k=1}^{n_u} \tau_{u,k}$$
$$+ \frac{1}{m_{11}} \tau_{wu}(t)$$

$$\dot{v} = -\frac{m_{11}}{m_{22}} ur - \frac{d_{22}}{m_{22}} v - \sum_{i=2}^{3} \frac{d_{vi}}{m_{22}} |v|^{i-1} v + \frac{1}{m_{22}} \tau_{wv}(t)$$

$$\dot{r} = \frac{m_{11} - m_{22}}{m_{33}} uv - \frac{d_{33}}{m_{33}} r - \sum_{i=2}^{3} \frac{d_{ri}}{m_{33}} |r|^{i-1} r + \frac{1}{m_{33}} \sum_{k=1}^{n_r} \tau_{r,k}$$
$$+ \frac{1}{m_{33}} \tau_{wr}(t) \tag{8.2}$$

where $\bar{x}, \bar{y}$ denote the position of the ship surface in X and Y directions, respectively. $\bar{\phi}$ denotes the heading angle of the ship. u, v and r are the surge, sway and yaw velocities, respectively. $\tau_{u,k}$ and $\tau_{r,k}$ represent the surge and yaw actuators, respectively, which are considered as actual control inputs; n_u and n_r are the number of actuators of surge and yaw force respectively. The failures that may occur on the kth actuators can be modeled as

$$\tau_{u,k}(t) = \delta_{k,h} \bar{\tau}_{u,k}(t) + \chi_{u,k,h}$$
$$\tau_{r,k}(t) = \varrho_{k,h} \bar{\tau}_{r,k}(t) + \chi_{r,k,h}, \quad t \in [T^s_{k,h}, T^e_{k,h}), \quad h = 1, 2, 3... \tag{8.3}$$

where $\delta_{k,h}$, $\varrho_{k,h}$, $T^s_{k,h}$ and $T^e_{k,h}$ are unknown constants with $0 \leq \delta_{k,h} < 1$, $0 \leq \varrho_{k,h} < 1$ and $0 \leq T^s_{k,1} < T^e_{k,1} \leq T^s_{k,2} < T^e_{k,2} \leq ... \leq \infty$, and $\chi_{u,k,h}$

and $\chi_{r,k,h}$ are unknown, piecewise continuous and bounded signals. Eq. (8.3) covers the following two types of failures:

1. $0 < \delta_{k,h} < 1$ or $0 < \varrho_{k,h} < 1$: In this case, the actuator is called Partial Loss of Effectiveness. The actuator gain, whose nominal value is 1, decreases to a value within the interval $(0, 1)$ and the actuator may suffer from an additive fault $\chi_{u,k,h}$ or $\chi_{r,k,h}$ simultaneously.

2. $\delta_{k,h} = 0$ or $\varrho_{k,h} = 0$: In this case, $\tau_{u,k} = \chi_{u,k,h}$ or $\tau_{r,k} = \chi_{r,k,h}$ and the actuator is called Total Loss of Effectiveness, i.e., the output $\tau_{u,k}$ or $\tau_{r,k}$ can no longer be influenced by the control input $\bar{\tau}_{u,k}$ or $\bar{\tau}_{r,k}$. For instance, it is encountered in practice that the actuator is stuck at some unknown value.

In Eq. (8.3), $T_{k,h}^s$ and $T_{k,h}^e$ denote the time instants when the hth failure on the kth actuator starts and ends respectively. If $T_{k,h+1}^s > T_{k,h}^e$, the actuator recovers its normal operation when the next failure takes place. If $T_{k,h+1}^s = T_{k,h}^e$, then it implies that the failure $\delta_{k,h}$ or $\varrho_{k,h}$ jumps to $\delta_{k,h+1}$ or $\varrho_{k,h+1}$ at time $T_{k,h}^e$ without recovering. In Eq. (8.3), h is not restricted to be finite and the total number failures is allowed to be infinite when time goes to infinity.

Introducing four piecewise continuous functions

$$\delta_k(t) = \begin{cases} \delta_{k,h} & if\ t \in [T_{k,h}^s, T_{k,h}^e), \\ 1 & if\ t \in [T_{k,h}^e, T_{k,h+1}^s) \end{cases}$$

$$\chi_{u,k}(t) = \begin{cases} \chi_{u,k,h} & if\ t \in [T_{k,h}^s, T_{k,h}^e), \\ 1 & if\ t \in [T_{k,h}^e, T_{k,h+1}^s) \end{cases}$$

$$\varrho_k(t) = \begin{cases} \varrho_{k,h} & if\ t \in [T_{k,h}^s, T_{k,h}^e), \\ 1 & if\ t \in [T_{k,h}^e, T_{k,h+1}^s) \end{cases}$$

$$\chi_{r,k}(t) = \begin{cases} \chi_{r,k,h} & if\ t \in [T_{k,h}^s, T_{k,h}^e), \\ 1 & if\ t \in [T_{k,h}^e, T_{k,h+1}^s) \end{cases}$$

, the kth actuator with the unknown failures (8.3) can be represented as

$$\tau_{u,k}(t) = \delta_k(t)\bar{\tau}_{u,k}(t) + \chi_{u,k}$$
$$\tau_{r,k}(t) = \varrho_k(t)\bar{\tau}_{r,k}(t) + \chi_{r,k}. \tag{8.4}$$

Since there is no force introduced in the sway direction, the system modeled by (8.1)-(8.2) is underactuated. The positive constants m_{jj} denote the ship inertia including added mass, and d_{jj}, d_{ui}, d_{vi}, d_{ri} for $2 \le i \le 3$, $1 \le j \le 3$ represent the hydrodynamic damping coefficients. Note that all these constants are assumed to be unknown. The time-varying terms, $\tau_{wu}(t)$, $\tau_{wv}(t)$ and $\tau_{wr}(t)$, are the environmental disturbances induced by the wave, wind and ocean-current.

We shall then propose a design scheme for the control inputs $\tau_{u,k}$ and $\tau_{r,k}$ to solve the path tracking control of the actuated surface ship with unknown system parameters and environmental disturbances. The control objective can be formally stated as follows.

• *Control Objective*: Design the control inputs $\tau_{u,k}$ and $\tau_{r,k}$ to force the ship, as modeled in (8.1)-(8.2), to follow a prescribed path set, which is denoted by $\Omega = (x_d(s), y_d(s))$ with s being a path parameter.

Lemma 8.1. *[95] For any scalars ϵ and $z > 0$, the following relationship holds:*

$$0 \leq |z| - \frac{z^2}{\sqrt{z^2 + \epsilon^2}} < \epsilon. \tag{8.5}$$

To achieve this objective, the following assumption is imposed.

Assumption 8.1 *(a) $x_d'(s)^2 + y_d'(s)^2 \geq \mu$ where μ is a positive constant, $x_d'(s) = \dfrac{\partial x_d(s)}{\partial s}$ and $y_d'(s) = \dfrac{\partial y_d(s)}{\partial s}$.*

(b) The minimum radius of the osculating circle of the path is larger than or equal to the minimum possible turning radius of the ships.

(c) The unknown parameters are in known convex compact sets.

(d) The disturbances satisfy

$$|\tau_{wu}(t)| < \tau_{wu\,\text{max}}, \quad |\tau_{wv}(t)| < \tau_{wv\,\text{max}}, \quad |\tau_{wr}(t)| < \tau_{wr\,\text{max}}$$

where $\tau_{wu\,\text{max}}$, $\tau_{wv\,\text{max}}$ and $\tau_{wr\,\text{max}}$ are also unknown positive constants.

(e) At any time instant, up to $k-1$ actuators are at the state of Total Loss of Effectiveness. For the Partial Loss of Effectiveness case, $\delta_k(t) \geq \mu_k > 0$ and $\varrho_k(t) \geq \mu_k > 0$, where μ_k is a unknown constant.

Remark 8.2. In Assumption 8.1, (a), (b) imply that the path Ω is regular with respect to the path parameter s. However, if the path is not regular, it can be divided into several different regular paths for consideration separatively.

8.3 Design of Adaptive Controllers

8.3.1 Ship Dynamics Transformation

In order to tackle the underactuated problem, a coordinate change is firstly performed as a preliminary step. The transverse function approach in [15] is adopted to introduce an "auxiliary manipulated variable" in addition to the actual control inputs.

- *Change of coordinates:*

$$\begin{bmatrix} x \\ y \end{bmatrix} = \begin{bmatrix} \bar{x} \\ \bar{y} \end{bmatrix} + R(\phi) \begin{bmatrix} f_1(\alpha) \\ f_2(\alpha) \end{bmatrix}$$

$$\phi = \bar{\phi} - f_3(\alpha) \tag{8.6}$$

where $R(\phi) = [\cos(\phi), -\sin(\phi); \sin(\phi), \cos(\phi)]$ and $f_l(\alpha)$ for $l = 1, 2, 3$ are three functions of α which will be determined later. Taking the derivatives of x, y and ϕ gives

$$\begin{bmatrix} \dot{x} \\ \dot{y} \end{bmatrix} = Q \begin{bmatrix} u \\ \dot{\alpha} \end{bmatrix} + \begin{bmatrix} -v \sin(\bar{\phi}) \\ v \cos(\bar{\phi}) \end{bmatrix}$$

$$+ R'(\phi) \begin{bmatrix} f_1(\alpha) \\ f_2(\alpha) \end{bmatrix} (r - f_3'(\alpha)\dot{\alpha}) \tag{8.7}$$

$$\dot{\phi} = r - f_3'\dot{\alpha} \tag{8.8}$$

where $\dot{\alpha}$ is deemed as an auxiliary manipulated variable, and

$$Q = \left[\begin{bmatrix} \cos(\bar{\phi}) \\ \sin(\bar{\phi}) \end{bmatrix}, \ R(\phi) \begin{bmatrix} f_1'(\alpha) \\ f_2'(\alpha) \end{bmatrix} \right],$$

$f_l' = \dfrac{\partial f_l(\alpha)}{\partial \alpha}$ for $l = 1, 2, 3$ and $R'(\phi) = \dfrac{\partial R(\phi)}{\partial \phi}$. $f_l(\alpha)$ are chosen as follows such that Q is ensured to be invertible for all $\bar{\phi} \in \Re$ and $\alpha \in \Re$.

$$f_1(\alpha) = \epsilon_1 \sin(\alpha) \frac{\sin(f_3)}{f_3}, \ f_2(\alpha) = \epsilon_1 \sin(\alpha) \frac{1 - \cos(f_3)}{f_3},$$

$$f_3(\alpha) = \epsilon_2 \cos(\alpha) \tag{8.9}$$

where ϵ_1 and ϵ_2 are constants selected to satisfy that

$$0 < \epsilon_1, \ 0 < \epsilon_2 < \frac{\pi}{2}$$

and it follows that

$$|f_1| < \epsilon_1, \ |f_2| < \epsilon_1, \ |f_3| < \epsilon_2$$

$$\det(Q) = \frac{\epsilon_1 \epsilon_2}{(\epsilon_2 \cos(\alpha))^2} (\cos(\epsilon_2 \cos(\alpha)) - 1)$$

$$\leq -\frac{\epsilon_1}{\epsilon_2}(1 - \cos(\epsilon_2))$$

$$< 0. \tag{8.10}$$

8.3.2 Controller Design

By replacing (8.1) with (8.7) and (8.8), the resulting system with (8.2), (8.7) and (8.8) is of strict feedback form. Therefore, we shall apply the backstepping technique [93] to design the control inputs τ_u and τ_r. Obviously, the design procedure can be divided into two steps. In the first step, the virtual controls u_d and r_d together with the "auxiliary manipulated variable" $\dot{\alpha}$ will be constructed to force the motion of the ship to follow the prescribed reference (x_d, y_d). In the second step, the actual control signals for τ_u and τ_r will be delivered such that u and r in (8.1) can approach the virtual controls u_d and r_d, respectively. Apart from these, the adaptive laws for the unknown system parameters will also be provided.

Step *1*: Define the path tracking errors as

$$x_e = x - x_d, \quad y_e = y - y_d, \quad \phi_e = \phi - \phi_d. \tag{8.11}$$

The reference angle ϕ_d is given as $\phi_d = \arctan(\dfrac{y_d'}{x_d'})$ where $x_d' = \dfrac{\partial x_d}{\partial s}$ and $y_d' = \dfrac{\partial y_d}{\partial s}$. Choose the Lyapunov function candidate in this step as

$$V_1 = \frac{1}{2} q_e^T q_e + \frac{1}{2} \phi_e^2 \tag{8.12}$$

where $q_e = [x_e, y_e]^T$. Then the derivative of V_1 can be computed as

$$\dot{V}_1 = q_e^T \left(Q \begin{bmatrix} u \\ \dot{\alpha} \end{bmatrix} + \begin{bmatrix} -v\sin(\bar{\phi}) \\ v\cos(\bar{\phi}) \end{bmatrix} + R'(\phi) \begin{bmatrix} f_1(\alpha) \\ f_2(\alpha) \end{bmatrix} \right.$$
$$\left. \times (r - f_3'(\alpha)\dot{\alpha}) - \dot{q}_d \right) + \phi_e(r - f_3'\dot{\alpha} - \dot{\phi}_d).$$

In the above expression, $q_d = [x_d, y_d]^T$ and $\dot{q}_d = [x_d'(s)\dot{s}, y_d'(s)\dot{s}]^T$, $\dot{\phi}_d = \dfrac{x_d'(s)y_d''(s) - x_d''(s)y_d'(s)}{x_d'(s)^2 + y_d'(s)^2}\dot{s}$, where $x_d''(s) = \dfrac{\partial^2 x_d}{\partial s^2}$, $y_d''(s) = \dfrac{\partial^2 y_d}{\partial s^2}$. $\dot{s}$ is selected to be

$$\dot{s} = \frac{u^\star(1 - \epsilon_3 e^{-\epsilon_4 t})e^{-\epsilon_5 \|q - q_d\|}}{\sqrt{x_d'^2 + y_d'^2}} \tag{8.13}$$

where $u^\star$ is a non-zero prescribed speed, $\epsilon_i > 0$ for $i = 3, 4, 5$, $\epsilon_3 < 1$ and $q = [x, y]^T$.

Remark 8.3. Note that in a path tracking problem, the tracking rate is freedom to be chosen; see $\dot{s}$ in (8.13). In this chapter, the guideline for $\dot{s}$ is determined mainly based on the following considerations. Parameters ϵ_3 and ϵ_4 guarantee that the 'reference ship' runs at a low speed from the beginning and speeds up slowly. This is desirable in practice because it avoids using a high gain control input when $\|q - q_d\|$ is large where . Also when the ship encounters a sudden turn, $\sqrt{x_d'^2 + y_d'^2}$ is large. Then the ship should slow down, which can be made possible by (8.13).

We then introduce two new error variables

$$u_e = u - u_d, r_e = r - r_d \tag{8.14}$$

where u_d and r_d are the virtual controls for u and r, respectively. u_d, r_d and $\dot{\alpha}$ are chosen as

$$\begin{bmatrix} u_d \\ \dot{\alpha} \end{bmatrix} = Q^{-1} \left(-k_1 q_e - \begin{bmatrix} -v \sin(\bar{\phi}) \\ v \cos(\bar{\phi}) \end{bmatrix} - R'(\phi) \begin{bmatrix} f_1(\alpha) \\ f_2(\alpha) \end{bmatrix} \right.$$

$$\left. \times (-k_2 \phi_e + \dot{\phi}_d) + \dot{q}_d \right) \tag{8.15}$$

$$r_d = -k_2 \phi_e + f_3' \dot{\alpha} + \dot{\phi}_d \tag{8.16}$$

where k_1 and k_2 are positive constants.

From (8.16), we obtain that

$$\dot{V}_1 = -k_1 q_e^T q_e - k_2 \phi_e^2 + \varrho_u u_e + \varrho_r r_e$$

where $\varrho_u = q_e^T Q[1, \ 0]^T$ and $\varrho_r = q_e^T R'(\phi)[f_1(\alpha), \ f_2(\alpha)]^T + \phi_e$.

Step 2: Taking derivatives of both sides of (8.14) yields

$$\dot{u}_e = \frac{m_{22}}{m_{11}} vr - \frac{d_{11}}{m_{11}} u - \sum_{i=2}^{3} \frac{d_{ui}}{m_{11}} |u|^{i-1} u + \frac{1}{m_{11}} \sum_{k=1}^{n_u} \delta_k(t) \bar{\tau}_{u,k}$$

$$+ \frac{1}{m_{11}} \bar{\tau}_{wu}(t) - \frac{\partial u_d}{\partial x_e} \dot{x}_e - \frac{\partial u_d}{\partial y_e} \dot{y}_e - \frac{\partial u_d}{\partial s} \dot{s} - \frac{\partial u_d}{\partial \phi} \dot{\phi}$$

$$- \frac{\partial u_d}{\partial \bar{\phi}} \dot{\bar{\phi}} - \frac{\partial u_d}{\partial \alpha} \dot{\alpha} - \frac{\partial u_d}{\partial v} \left(-\frac{m_{11}}{m_{22}} ur - \frac{d_{22}}{m_{22}} v \right.$$

$$\left. - \sum_{i=2}^{3} \frac{d_{vi}}{m_{22}} |v|^{i-1} v + \frac{1}{m_{22}} \tau_{wv}(t) \right)$$

$$- \frac{\partial u_d}{\partial x_d'} \ddot{x}_d - \frac{\partial u_d}{\partial y_d'} \ddot{y}_d - \frac{\partial u_d}{\partial \phi_d'} \ddot{\phi}_d$$

$$\dot{r}_e = \frac{m_{11} - m_{22}}{m_{33}} uv - \frac{d_{33}}{m_{33}} r - \sum_{i=2}^{3} \frac{d_{ri}}{m_{33}} |r|^{i-1} r + \frac{1}{m_{33}} \sum_{k=1}^{n_r} \varrho_k(t) \bar{\tau}_{r,k}$$

$$+ \frac{1}{m_{33}} \bar{\tau}_{wr}(t) - \frac{\partial r_d}{\partial x_e} \dot{x}_e - \frac{\partial r_d}{\partial y_e} \dot{y}_e - \frac{\partial r_d}{\partial s} \dot{s} - \frac{\partial r_d}{\partial \phi} \dot{\phi}$$

$$- \frac{\partial r_d}{\partial \dot{\phi}} \ddot{\phi} - \frac{\partial r_d}{\partial \alpha} \dot{\alpha} - \frac{\partial r_d}{\partial \dot{\alpha}} \left(\frac{\partial \dot{\alpha}}{\partial x_e} \dot{x}_e + \frac{\partial \dot{\alpha}}{\partial y_e} \dot{y}_e + \frac{\partial \dot{\alpha}}{\partial s} \dot{s} + \frac{\partial \dot{\alpha}}{\partial \phi} \dot{\phi} \right.$$

$$+ \frac{\partial \dot{\alpha}}{\partial \dot{\phi}} \ddot{\phi} \right) - \frac{\partial \dot{\alpha}}{\partial v} \left(- \frac{m_{11}}{m_{22}} ur - \frac{d_{22}}{m_{22}} v - \sum_{i=2}^{3} \frac{d_{vi}}{m_{22}} |v|^{i-1} v \right.$$

$$+ \frac{1}{m_{22}} \tau_{wv}(t) \right) - \frac{\partial r_d}{\partial x_d'} \ddot{x}_d - \frac{\partial r_d}{\partial y_d'} \ddot{y}_d - \frac{\partial r_d}{\partial \phi_d'} \ddot{\phi}_d,$$

where $\bar{\tau}_{wu}(t) = \tau_{wu}(t) + \sum_{k=1}^{n_u} \chi_{u,k}$ and $\bar{\tau}_{wr}(t) = \tau_{wr}(t) + \sum_{k=1}^{n_r} \chi_{r,k}$. From Assumptions e, it can be checked that $\sum_{k=1}^{n_u} \delta_k(t) > \mu_k > 0$, thus Thus we have $\inf_{t \geq 0} \sum_{k=1}^{n_u} \delta_k(t) > 0$. To compensate for the unknown actuator failures, define

$$l_u = \inf_{t>0} \sum_{k=1}^{n_u} \delta_k(t), p_u = \frac{1}{l_u}$$

$$l_r = \inf_{t>0} \sum_{k=1}^{n_r} \varrho_k(t), p_r = \frac{1}{l_r}. \tag{8.17}$$

We choose the Lyapunov function candidate in this step as

$$V_2 = V_1 + \frac{m_{11}}{2} u_e^2 + \frac{m_{33}}{2} r_e^2 + \frac{1}{2} \sum_{i=1}^{3} \tilde{\theta}_i^T \Gamma_i^{-1} \tilde{\theta}_i \tag{8.18}$$

where $\Gamma_i = \text{diag}(\delta_{ij})$ for $i = 1, 2, 3$ are positive definite matrices with appropriate dimensions. $\tilde{\theta}_i = \theta_i - \hat{\theta}_i$ denote the estimation errors. $\hat{\theta}_i$ is the estimate designed for θ_i, where

$$\theta_1 = \left[m_{22}, d_u, d_{u2}, d_{u3}, m_{11}, \frac{m_{11}^2}{m_{22}}, \frac{d_v m_{11}}{m_{22}}, \frac{d_{v2} m_{11}}{m_{22}}, \right.$$

$$\left. \frac{d_{v3} m_{11}}{m_{22}} \right]$$

$$\theta_2 = \left[(m_{11} - m_{22}), d_r, d_{r2}, d_{r3}, m_{33}, \frac{m_{11} m_{33}}{m_{22}}, \frac{d_v m_{33}}{m_{22}}, \right.$$

$$\left. \frac{d_{v2} m_{33}}{m_{22}}, \frac{d_{v3} m_{33}}{m_{22}} \right]$$

$$\theta_3 = \left[\tau_{wu \max}, \frac{m_{11}}{m_{22}} \tau_{wv \max}, \tau_{wr \max}, \frac{m_{33}}{m_{22}} \tau_{wv \max} \right].$$

Without canceling the useful damping term, the actual controls for τ_u and τ_r are designed as

$$\alpha_u = -k_3 u_e - \hat{\theta}_1^T \beta_1 + \varrho_u - \hat{\theta}_{31} \tanh(\frac{u_e \hat{\theta}_{31}}{\varepsilon_1})$$

$$- \hat{\theta}_{32} \tanh(\frac{\partial u_d}{\partial v} \frac{u_e \hat{\theta}_{32}}{\varepsilon_2})$$

$$\alpha_r = -k_4 r_e - \hat{\theta}_2^T \beta_2 + \varrho_r - \hat{\theta}_{33} \tanh(\frac{r_e \hat{\theta}_{33}}{\varepsilon_3})$$

$$- \hat{\theta}_{34} \tanh(\frac{\partial r_d}{\partial v} \frac{r_e \hat{\theta}_{34}}{\varepsilon_4})$$

$$\bar{\tau}_{u,k} = -\frac{u_e \hat{p}_u^2 \alpha_u^2}{\sqrt{u_e^2 \hat{p}_u^2 \alpha_u^2 + \epsilon(t)^2}}, k = 1, ..., n_u$$

$$\bar{\tau}_{r,m} = -\frac{r_e \hat{p}_r^2 \alpha_r^2}{\sqrt{r_e^2 \hat{p}_r^2 \alpha_r^2 + \epsilon(t)^2}}, m = 1, ..., n_r \qquad (8.19)$$

where k_3, k_4, ε_i, $1 \le i \le 4$ are positive constants, $\epsilon(t) = \iota e^{-\nu t}$ with ι and ν being positive constants, and

$$\beta_1 = \left[vr, -u_d, -|u|u_d, -u^2 u_d, \gamma_1, \frac{\partial u_d}{\partial v} ur, \frac{\partial u_d}{\partial v} v, \right.$$
$$\left. \frac{\partial u_d}{\partial v}|v|v, \frac{\partial u_d}{\partial v} v^3 \right]$$

$$\beta_2 = \left[uv, -r_d, -|r|r_d, -r^2 r_d, \gamma_2, \frac{\partial r_d}{\partial v} ur, \frac{\partial r_d}{\partial v} v, \right.$$
$$\left. \frac{\partial r_d}{\partial v}|v|v, \frac{\partial r_d}{\partial v} v^3 \right]$$

$$\gamma_1 = -\frac{\partial u_d}{\partial x_e}\dot{x}_e - \frac{\partial u_d}{\partial y_e}\dot{y}_e - \frac{\partial u_d}{\partial s}\dot{s} - \frac{\partial u_d}{\partial \phi}\dot{\phi} - \frac{\partial u_d}{\partial \bar{\phi}}\dot{\bar{\phi}}$$
$$- \frac{\partial u_d}{\partial \alpha}\dot{\alpha} - \frac{\partial u_d}{\partial x_d'}\ddot{x}_d - \frac{\partial u_d}{\partial y_d'}\ddot{y}_d - \frac{\partial u_d}{\partial \phi_d'}\ddot{\phi}_d$$

$$\gamma_2 = -\frac{\partial r_d}{\partial x_e}\dot{x}_e - \frac{\partial r_d}{\partial y_e}\dot{y}_e - \frac{\partial r_d}{\partial s}\dot{s} - \frac{\partial r_d}{\partial \phi}\dot{\phi} - \frac{\partial r_d}{\partial \bar{\phi}}\dot{\bar{\phi}}$$
$$- \frac{\partial r_d}{\partial \alpha}\dot{\alpha} - \frac{\partial r_d}{\partial \dot{\alpha}} \left(\frac{\partial \dot{\alpha}}{\partial x_e}\dot{x}_e + \frac{\partial \dot{\alpha}}{\partial y_e}\dot{y}_e + \frac{\partial \dot{\alpha}}{\partial s}\dot{s} + \frac{\partial \dot{\alpha}}{\partial \phi}\dot{\phi} \right.$$
$$\left. + \frac{\partial \dot{\alpha}}{\partial \bar{\phi}}\dot{\bar{\phi}} \right) - \frac{\partial r_d}{\partial x_d'}\ddot{x}_d - \frac{\partial r_d}{\partial y_d'}\ddot{y}_d - \frac{\partial r_d}{\partial \phi_d'}\ddot{\phi}_d.$$

The adaptive laws are chosen as

$$\dot{\hat{\theta}}_{1j} = \delta_{1j}\mathrm{Proj}(u_e\beta_{1j}, \hat{\theta}_{1j}), \quad 1 \le j \le 9,$$

$$\dot{\hat{\theta}}_{2j} = \delta_{2j}\mathrm{Proj}(r_e\beta_{2j}, \hat{\theta}_{2j}), \quad 1 \le j \le 9,$$

$$\dot{\hat{\theta}}_{31} = \delta_{31}\mathrm{Proj}(|u_e|, \hat{\theta}_{31})$$

$$\dot{\hat{\theta}}_{32} = \delta_{32}\mathrm{Proj}(|u_e\frac{\partial u_d}{\partial v}|, \hat{\theta}_{32})$$

$$\dot{\hat{\theta}}_{33} = \delta_{33}\mathrm{Proj}(|r_e|, \hat{\theta}_{33})$$

$$\dot{\hat{\theta}}_{34} = \delta_{34}\mathrm{Proj}(|r_e\frac{\partial r_d}{\partial v}|, \hat{\theta}_{34}) \tag{8.20}$$

where $\dot{\hat{\theta}}_{ij}$ denotes the jth entry of $\dot{\hat{\theta}}_i$. δ_{1j}, δ_{2j} and δ_{3i}, $q \le i \le 4$, $1 \le j \le 9$ are positive constants. β_{1j} and β_{2j} are the jth element of β_1 and β_2 respectively. $\hat{p}_u$ and $\hat{p}_r$ are updated according to

$$\dot{\hat{p}}_u = \gamma_u u_e \alpha_u$$

$$\dot{\hat{p}}_r = \gamma_r r_e \alpha_r \tag{8.21}$$

where γ_u and γ_r are positive constants. With the aid of Lemma 8.1, we have

$$\sum_{k=1}^{n_u} u_e \delta_k \bar{\tau}_{u,k} = -\sum_{k=1}^{n_u} \delta_k \frac{u_e^2 \hat{p}_u^2 \alpha_u^2}{\sqrt{u_e^2 \hat{p}_u^2 \alpha_u^2 + \epsilon(t)^2}}$$

$$\le -\frac{l_u u_e^2 \hat{p}_u^2 \alpha_u^2}{\sqrt{u_e^2 \hat{p}_u^2 \alpha_u^2 + \epsilon(t)^2}}$$

$$\le l_u\epsilon(t) - u_e l_u \hat{p}_u \alpha_u \tag{8.22}$$

and

$$\sum_{k=1}^{n_r} r_e \delta_k \bar{\tau}_{r,k} = -\sum_{k=1}^{n_r} \delta_k \frac{r_e^2 \hat{p}_r^2 \alpha_r^2}{\sqrt{r_e^2 \hat{p}_r^2 \alpha_r^2 + \epsilon(t)^2}}$$

$$\le -\frac{l_r r_e^2 \hat{p}_r^2 \alpha_r^2}{\sqrt{r_e^2 \hat{p}_r^2 \alpha_r^2 + \epsilon(t)^2}}$$

$$\le l_r\epsilon(t) - r_e l_r \hat{p}_r \alpha_r. \tag{8.23}$$

The operator Proj is a Lipschitz continuous projection algorithm [94] defined as follows:

$$\mathrm{Proj}(a, \hat{b}) = \begin{cases} a & \text{if } \mu(\hat{b}) \le 0 \\ a & \text{if } \mu(\hat{b}) \ge 0 \text{ and } \mu'(\hat{b})a \le 0 \\ (1 - \mu(\hat{b}))a & \text{if } \mu(\hat{b}) > 0 \text{ and } \mu'(\hat{b})a > 0 \end{cases} \tag{8.24}$$

where $\mu(\hat{b}) = \dfrac{\hat{b}^2 - b_M^2}{\epsilon^2 + 2\epsilon b_M}$, $\mu'(\hat{b}) = \dfrac{\partial\mu(\hat{b})}{\hat{b}}$, ϵ is an arbitrarily small positive constant, b_M is a positive constant satisfying $|b| < b_M$. The following results are then obtained.

Lemma 8.4. *The projection has the following properties if $|\hat{b}(t_0)| \leq b_M$:*

(1) $|\hat{b}(t)| \leq b_M + \epsilon$, $\forall 0 \leq t_0 \leq t < \infty$.

(2) $Proj(a, \hat{b})$ is Lipschitz continuous.

(3) $|Proj(a, \hat{b})| \leq |a|$

(4) $\tilde{b} Proj(a, \hat{b}) \geq \tilde{b} a$ where $\tilde{b} = b - \hat{b}$.

Proof: See [93].   □

8.3.3 Stability Analysis

Based on Lemma 8.1, then by differentiating both sides of (8.18), and using (8.19), (8.20) and (8.24), we have

$$\dot{V}_2(t) \leq -k_1 q_e^T q_e - k_2 \phi_e^2 - (k_3 + d_u)u_e^2 - (k_4 + d_r)r_e^2$$
$$+ 0.2785 \sum_{i=1}^{4} \varepsilon_i + (l_u + l_r)\epsilon(t) \tag{8.25}$$

which means

$$\dot{V}_2 \leq -\rho_1 V_2 + \rho_2 \tag{8.26}$$

where

$$\rho_1 = \min\left(1, 2k_1, 2k_2, \frac{2(k_3 + d_u)}{m_{11}}, \frac{2(k_4 + d_r)}{m_{33}}\right)$$

$$\rho_2 = \frac{1}{2}\sum_{i=1}^{3} \tilde{\theta}_i^T \Gamma_i^{-1} \tilde{\theta}_i + 0.2785 \sum_{i=1}^{4} \varepsilon_i + (l_u + l_r)\epsilon(t).$$

From (8.26), we get

$$V_2(t) \leq V_2(t_0)e^{-\rho_1(t-t_0)} + \frac{\rho_2}{\rho_1}. \tag{8.27}$$

From (8.27), it is obvious that q_e, ϕ_e, u_e and r_e are bounded with known upper-bounds.

From (8.15), (8.16), u_d and r_d can be rewritten as

$$u_d = \xi_u + \omega_u v \tag{8.28}$$
$$r_d = \xi_r + \omega_r v. \tag{8.29}$$

where

$$\omega_u = [1, 0]Q^{-1}[\sin(\bar\phi), 0]^T \tag{8.30}$$

$$\omega_r = f_3'[0, 1]Q^{-1}[0, -\cos\bar\phi]^T \tag{8.31}$$

Then, the sway velocity dynamics in (8.2) can be rewritten as

$$\dot v = \varpi_1 v + \varpi_2 v^2 - \frac{d_{v2}}{m_{22}}|v|v - \frac{d_{v3}}{m_{22}}v^3 + \varpi_3 \tag{8.32}$$

where

$$\varpi_1 = -\frac{m_{11}}{m_{22}}(\xi_u\omega_r + \xi_r\omega_u + r_e\omega_u + u_e\omega_r) - \frac{d_{11}}{m_{11}}$$

$$\varpi_2 = -\frac{m_{11}}{m_{22}}\omega_u\omega_r$$

$$\varpi_3 = -\frac{m_{11}}{m_{22}}(\xi_u\xi_r + u_e r_e + r_e\xi_u + u_e\xi_r) + \frac{1}{m_{22}}\tau_{wv}(t).$$

Since q_e, ϕ_e, u_e and r_e are bounded with known upper-bounds, it is easy to check that ϖ_1, ϖ_2 and ϖ_3 are bounded with known bounds. To show that v is bounded, the following Lyapunov function is defined

$$V_3 = \frac{1}{2}v^2, \tag{8.33}$$

whose derivative along the solution of (8.33) satisfies

$$\dot V_3 \le -\left(\frac{d_{v3}}{m_{22}} - \varpi_2^M\eta\right)v^4 + \left(\varpi_1^M + \frac{\varpi_2^M + \varpi_3^M}{4\eta}\right)v^2 + \varpi_3^M\eta \tag{8.34}$$

where ϖ_i^M for $i = 1, 3$ are the upper-bounds of ϖ_i, ϖ_2^M is the upper-bound of $\varpi_2 + \frac{d_{v2}}{n_{22}}$. Thus we can choose a η such that $\frac{d_{v3}}{m_{22}} - \varpi_2^M\eta > \eta^\star > 0$, where $\eta^\star$ is a positive constant. Then from [66], it follows that (8.34) guarantees a finite upper-bound of the sway velocity v.

Based on the above analysis, the following theorem can be drawn.

Theorem 8.5. *Consider the underactuated ship system (8.1) and (8.2), with the controller (8.19) and parameter update laws (8.20) under Assumption 8.1. All the signals in the closed loop system are bounded, and the reference trajectories $x_d(s)$ and $y_d(s)$ can be tracked with bounded and arbitrarily small tracking errors.*

Proof: Due to the projection operation, $\tilde\theta_i$, $i = 1, 2, 3$ are bounded. Thus from (8.27), all signals contained in V_2 are bounded. Hence, x_e, y_e and ϕ_e are bounded. From (8.15) and (8.16), it is easy to check that u_d, r_d and $\dot\alpha$ are bounded. Thus u, v and r are bounded. From (8.19), the boundedness of τ_u

and τ_r is concluded.

From (8.6) and (8.9) it is easy to show that

$$\|(x - \bar{x}, y - \bar{y})\| \leq \sqrt{2\epsilon_1^2}, \quad |\phi - \bar{\phi}| \leq \epsilon_2. \tag{8.35}$$

Finally from (8.6), the tracking errors $\bar{x} - x_d$ and $\bar{y} - y_d$ can be shown to satisfy

$$|\bar{x} - x_d| \leq |\bar{x} - x| + |x_e|, \quad |\bar{y} - y_d| \leq |\bar{y} - y| + |y_e|. \tag{8.36}$$

Thus the tracking errors are bounded. On the other hand, since ϵ_1, ϵ_2, ε_i, $i = 1, 2, 3, 4$ can be made arbitrarily small. By adjusting Γ_j^{-1}, $j = 1, 2, 3$, the term ρ_2 in (8.26) can also be made arbitrarily small. Then from (8.27), x_e and y_e can be made arbitrarily small. Therefore, the tracking errors can also be made arbitrarily small. □

8.4 Simulation Results

In this section, we present our simulation results obtained using MATLAB. The parameters of the ship are $m_1 = 120 \times 10^3$kg, $m_2 = 60 \times 10^4$kg, $d_{u1} = 20 \times 10^3$kg, $d_{v1} = 140 \times 10^3$kg, $d_{r1} = 600 \times 10^3$kg, $d_{u2} = 0.2d_{u1}$, $d_{u3} = 0.1d_{u1}$, $d_{v2} = 0.2d_{v1}$, $d_{v3} = 0.1d_{v1}$, $d_{r2} = 0.2d_{r1}$, $d_{r3} = 0.1d_{r1}$. All the parameters are assumed to be unknown for controller design. The maximum and minimum values are set to fluctuate 30% above and below their true values. The bounds of the external disturbances are assumed to be 100. The design parameters are chosen as follows: $k_1 = 2$, $k_2 = 3$, $k_3 = 8$, $k_4 = 8$, $\epsilon_1 = 0.2$, $\epsilon_2 = 0.2$, $\epsilon_3 = 0.1$, $\epsilon_4 = 4$, $\epsilon_5 = 0.01$, $u^\star = 1.3$, $\varepsilon_i = 0.5$, $i = 1, 2, 3, 4$. The reference trajectories are taken as $q_{od} = [s, 10\sin(0.1s)]^T$. The initial conditions are set as $\{\bar{x}(0), \bar{y}(0), \bar{\phi}(0), u(0), v(0), r(0)\} = \{-5, 5, 3, 0, 0, 0\}$. All initial values of parameter estimates are taken to be 80% of their assumed true values. We assume in surge direction there are two actuators, and there is no actuator failure in yaw direction. $\delta_1(t) = 1$ for $t \leq 50s$ and $\delta_1(t) = 0$ for $50 < t \leq 100s$. $\delta_2(t) = 0$ for $t \leq 50s$ and $\delta_2(t) = 1$ for $50 < t \leq 100s$. The results on the ship position and tracking errors evolving with time are shown in Fig.8.2 and Fig.8.3, respectively. It can be seen that the references can be tracked with arbitrarily small tracking errors. Fig.8.4, Fig.8.5 and Fig.8.6 show the parameter estimator of θ_1, θ_2 and θ_3. Fig.8.7- Fig.8.9 show the surge, sway and yaw velocities of the ship respectively.

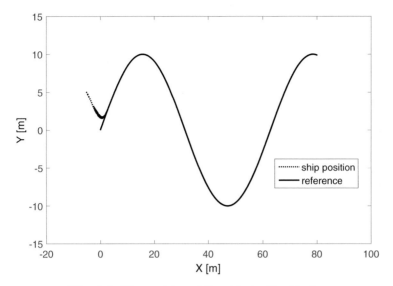

Fig. 8.2. The position of the ship in $X - Y$ plane.

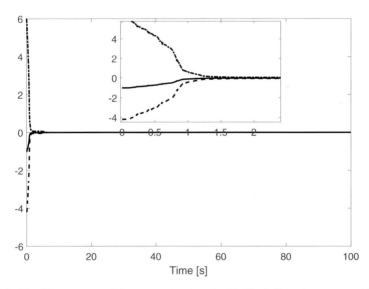

Fig. 8.3. Tracking errors with respect to time in X, Y, ϕ directions, respectively.

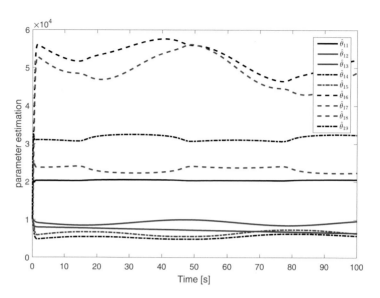

Fig. 8.4. Parameter estimator of θ_1.

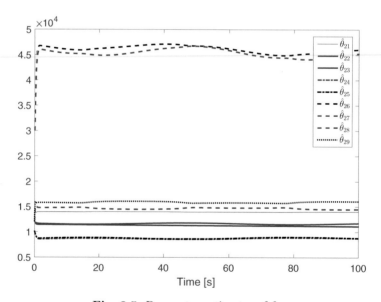

Fig. 8.5. Parameter estimator of θ_2.

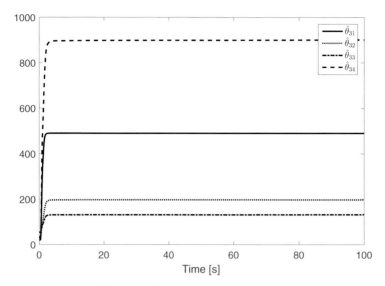

Fig. 8.6. Parameter estimator of θ_3.

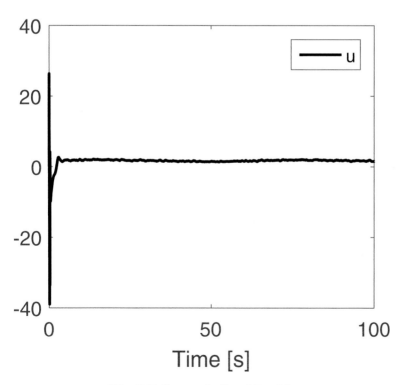

Fig. 8.7. Surge velocity of the ship.

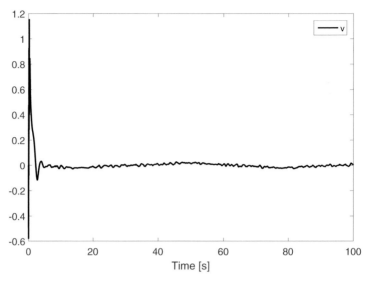

Fig. 8.8. Sway velocity of the ship.

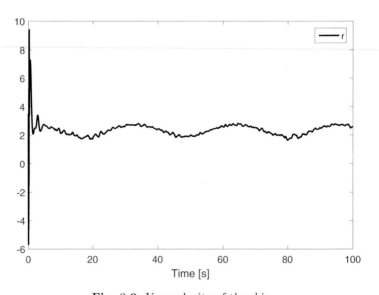

Fig. 8.9. Yaw velocity of the ship.

8.5 Conclusions

In this chapter, path tracking fault-tolerant control of underactuated ships with unknown system parameters and external disturbances induced by wave, wind and ocean current is investigated. By using the transverse function approach, we develop a globally stable adaptive controller to achieve arbitrarily small tracking errors. It should be noted that the transverse function approach can also be applied to other underactuated mechanical systems such as nonholonomic mobile robots, underwater vehicles and VTOL aircrafts etc.

9

Adaptive Fault-Tolerant Control of Underactuated Ships without Actuator Redundancy

This chapter studies the universal control design for global tracking control problem of a surface vessel when sway force suffers from nonperiodical actuator failure and the dynamics of the surface vessel switches between an underactuated system and an actuated system. A universal control scheme is proposed in this chapter such that it could be applied in both cases. Simulation results validate our proposed control methodology.

9.1 Introduction

A system is underactuated if there are a fewer number of independent actuators than the degrees of freedom. In reality, a lot of nonlinear mechanical systems belong to this type. In recent years many novel nonlinear design techniques have been developed for this control problem, mainly based on a time-varying feedback and discontinuous state-feedback control scheme. For surface vessel when the sway force is taken away with only surge force and yaw force, researchers proposed a number of control schemes. In [68], a discontinuous state-feedback control law is proposed using σ-process to exponentially stabilize the underactuated surface vessel to the origin. In [5], a discontinuous time-varying feedback controller for a nonholonomic system and this scheme is applied to an underactuated surface vessel. In [69] some local exponential stabilization results are developed based on a time-varying homogeneous control approach. In [70] an application of this homogeneous system theory to underactuated surface vessel was considered. In [71] some dynamic feedback results of stabilization were developed by transforming the underactuated surface vessel kinematics and dynamics into the so-called skew form. However, in some cases, the sway force of the surface vessel is necessary and it must be used occasionally. Thus it is naturally to consider a universal control scheme for the surface vessel such that when sway force is taken on or off, the control scheme could be applied in both cases.

In this chapter we consider designing a universal controller for the global tracking control problem of a surface vessel when sway force suffers from nonperiodical actuator failure and when the dynamics of the surface vessel switches between an underactuated system and an actuated system by using the Lyapunov's direct approach. We advocate Lyapunov's direct approach because it has become a standard method for nonlinear control design. This methodology could also be extended to other MIMO nonlinear systems whose actuators suffer from partial or total failure.

9.2 Problem Statement

The motion of the surface vessel could be described as the following differential equations:

$$
\begin{bmatrix} \dot{x} \\ \dot{y} \\ \dot{\phi} \end{bmatrix} = \begin{bmatrix} \cos\phi & -\sin\phi & 0 \\ \sin\phi & \cos\phi & 0 \\ 0 & 0 & 1 \end{bmatrix} \begin{bmatrix} u \\ v \\ r \end{bmatrix}
\tag{9.1}
$$

$$
\begin{aligned}
\dot{u} &= \frac{m_{22}}{m_{11}} vr - \frac{d_{11}}{m_{11}} u + \frac{1}{m_{11}} \tau_u \\
\dot{v} &= -\frac{m_{11}}{m_{22}} ur - \frac{d_{22}}{m_{22}} v + \frac{P(t)}{m_{22}} \tau_v \\
\dot{r} &= \frac{m_{11} - m_{22}}{m_{33}} uv - \frac{d_{33}}{m_{33}} r + \frac{1}{m_{33}} \tau_r
\end{aligned}
\tag{9.2}
$$

where (x, y) denotes the coordinate of the a surface vessel, ϕ is the heading angle, u, v and r denote the velocities in surge, sway and yaw respectively. τ_u, τ_v and τ_r are surge force, sway force and yaw force respectively, which are considered as input. It has $P(t) = 0$ or 1. When $P(t) = 1$, the vessel is fully actuated while when $P(t) = 0$ the vessel is underactuated. Fig.9.1 is the schematic figure of $P(t)$ and $Q(t)$ where we define $Q(t) = 1 - P(t)$. $P(t)$ switches to 0 from 1 at T_d, which means the system loses side force and becomes a underactuated system from a fully actuated system, and we call it a down case. Similarly we can define an up case. Without losing generality, we assume the initial value of $P(t)$ is 1 and it experiences n up cases and $n + 1$ down cases and eventually $P(t) = 0$.

The desired trajectory is generated by the dynamic equations of a virtual vessel:

$$
\begin{bmatrix} \dot{x}_d \\ \dot{y}_d \\ \dot{\phi}_d \end{bmatrix} = \begin{bmatrix} \cos\phi_d & -\sin\phi_d & 0 \\ \sin\phi_d & \cos\phi_d & 0 \\ 0 & 0 & 1 \end{bmatrix} \begin{bmatrix} u_d \\ v_d \\ r_d \end{bmatrix}
\tag{9.3}
$$

$$
\dot{v}_d = -\frac{m_{11}}{m_{22}} u_d r_d - \frac{d_{22}}{m_{22}} v_d
\tag{9.4}
$$

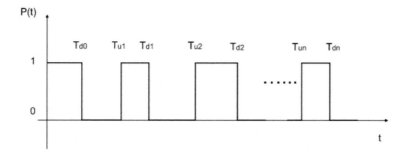

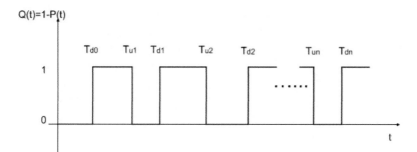

Fig. 9.1. A possible example of P(t) and Q(t).

where (x_d, y_d, ϕ_d) denotes the desired position and orientation of a virtual vessel; (u_d, v_d, r_d) denotes the desired velocities. The following assumption is imposed.

Assumption 9.1 *The reference velocities u_d and r_d and their derivatives $\dot{u}_d$, $\dot{r}_d$ are bounded.*

9.3 Controller Design

Firstly a coordinate transformation is made as follows

$$\begin{bmatrix} z_1 \\ z_2 \\ z_3 \end{bmatrix} = \begin{bmatrix} \cos\phi & \sin\phi & 0 \\ -\sin\phi & \cos\phi & 0 \\ 0 & 0 & 1 \end{bmatrix} \begin{bmatrix} x \\ y \\ \phi \end{bmatrix} \tag{9.5}$$

and define tracking error variables $z_{ie} = z_i - z_{id}$ with $i = 1, 2, 3$, $u_e = u - u_d$, $v_e = v - v_d$, $r_e = r - r_d$, and we get the following differential equations:

$$\dot{z}_{1e} = u_e + z_{2e}r_d + z_2 r_e$$

$$\dot{z}_{2e} = v_e - z_{1e}r_d - z_1 r_e$$

$$\dot{z}_{3e} = r_e$$

$$\dot{u}_e = \frac{m_{22}}{m_{11}}vr - \frac{d_{11}}{m_{11}}u + \frac{1}{m_{11}}\tau_u - \dot{u}_d$$

$$\dot{v}_e = -\frac{m_{11}}{m_{22}}u_e r_d - \frac{d_{22}}{m_{22}}v_e - \frac{m_{11}}{m_{22}}u r_e + \frac{P(t)}{m_{22}}\tau_v$$

$$\dot{r}_e = \frac{m_{11} - m_{22}}{m_{33}}uv - \frac{d_{33}}{m_{33}}r + \frac{1}{m_{33}}\tau_r - \dot{r}_d. \tag{9.6}$$

Clearly the tracking control problem is converted to the stabilization of the system (9.6) to the origin with control variables τ_u, τ_v and τ_r.

The actuator loss of τ_v should be recovered by adding compensations on τ_u and τ_r when $P(t) = 0$. To accomplish this, $Q(t)$ should be approximated by a smooth function. First define $\Psi(x)$ as follows

$$\Psi(x) = \begin{cases} 0 & x = 0 \\ 1 & x \neq 0. \end{cases} \tag{9.7}$$

The first $S-$function for the first down case is defined as

$$Sd_0(t) = 1 - e^{-\beta \int_0^t \int_0^t Q(u)du} \tag{9.8}$$

where β is a positive constant. The ith $S-$functions for ith up case and down case are defined as

$$Su_i(t) = 1 - e^{-\beta \int_0^t \int_0^t [\Psi(Sd_{i-1}(u))P(u)]du} \tag{9.9}$$

and

$$Sd_i(t) = 1 - e^{-\beta \int_0^t \int_0^t [\Psi(Su_i(u))Q(u)]du}. \tag{9.10}$$

Eventually we define a $S-$type function

$$S(t) = Sd_0(t) + \sum_{j=1}^{n} [-Su_j(t) + Sd_j(t)]. \tag{9.11}$$

A schematic figure of $S(t)$ is shown in Fig.9.2. It is a smooth function with a continuous and bounded derivative. Introduce a new variable $\bar{v}_e = v_e + \mu S(t)z_{2e}$, where μ is a positive constant, and we get

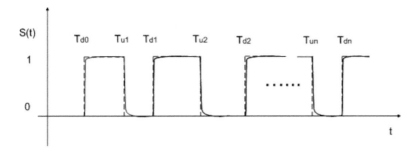

Fig. 9.2. A schematic figure of S(t).

$$\dot{z}_{1e} = u_e + z_{2e}r_d + z_2 r_e$$

$$\dot{z}_{2e} = \overline{v}_e - \mu S z_{2e} - z_{1e}r_d - z_1 r_e$$

$$\dot{z}_{3e} = r_e$$

$$\dot{u}_e = \frac{m_{22}}{m_{11}}vr - \frac{d_{11}}{m_{11}}u + \frac{1}{m_{11}}\tau_u - \dot{u}_d$$

$$\dot{\overline{v}}_e = -\frac{m_{11}}{m_{22}}u_e r_d - \frac{d_{22}}{m_{22}}\overline{v}_e + \frac{\mu S d_{22}}{m_{22}}z_{2e} - \frac{m_{11}}{m_{22}}u r_e + \mu S \dot{z}_{2e} + \mu \dot{S} z_{2e} + \frac{P(t)}{m_{22}}\tau_v$$

$$\dot{r}_e = \frac{m_{11} - m_{22}}{m_{33}}uv - \frac{d_{33}}{m_{33}}r + \frac{1}{m_{33}}\tau_r - \dot{r}_d. \qquad (9.12)$$

Firstly, in (z_{1e}, z_{2e}, z_{3e}) subsystem, let $u_e, \overline{v}_e$ and r_e be the virtual controllers, and define $z_4 = u_e - \alpha_1$, $z_5 = \overline{v}_e - \alpha_2$ and $z_6 = r_e - \alpha_3$, we get

$$\dot{z}_{1e} = z_4 + \alpha_1 + z_{2e}r_d + z_2 r_e$$

$$\dot{z}_{2e} = z_5 + \alpha_2 - \mu S z_{2e} - z_{1e}r_d - z_1 r_e$$

$$\dot{z}_{3e} = z_6 + \alpha_3.$$

Define a Lyapunov function $V_0 = \dfrac{\lambda_1}{2}z_{1e}{}^2 + \dfrac{\lambda_1}{2}z_{2e}{}^2$, where λ_1 is a positive constant, and we get the derivative of V_0 as

$$\dot{V}_0 = \lambda_1 z_{1e}(z_4 + \alpha_1 + z_2(z_6 + \alpha_3)) + \lambda_1 z_{2e}(z_5 + \alpha_2 - \mu S z_{2e} - z_1(z_6 + \alpha_3)).$$

The virtual control variables α_1 and α_2 are designed as:

$$\begin{cases} \alpha_1 = -\varepsilon_1 z_{1e} \\ \alpha_2 = -\mu(1 - S(t))z_{2e} \end{cases} \qquad (9.13)$$

where ε_1 is a positive constant, and we get

$$\dot{V}_0 = -\varepsilon_1 \lambda_1 z_{1e}{}^2 - \mu \lambda_1 z_{2e}{}^2 + \lambda_1 z_{1e}(z_4 + z_2 z_6) + \\ \lambda_1 z_{2e}(z_5 - z_1 z_6) + \alpha_3(\lambda_1 z_{1e} z_2 - \lambda_1 z_1 z_{2e}).$$

Define a new Lyapunov function $V_1 = V_0 + \dfrac{1}{2}z_4^2 + \dfrac{\lambda_2}{2}z_5^2$, where λ_2 is a positive constant, and we get the derivative of V_1 as:

$$\dot{V}_1 = \dot{V}_0 + z_4\left[\frac{m_{22}}{m_{11}}vr - \frac{d_{11}}{m_{11}}u + \frac{1}{m_{11}}\tau_u - \dot{u}_d - \dot{\alpha}_1\right]$$
$$+ \lambda_2 z_5\left\{-\frac{m_{11}}{m_{22}}(z_4 + \alpha_1)r_d - \frac{d_{22}}{m_{22}}(z_5 + \alpha_2) + \frac{\mu S d_{22}}{m_{22}}z_{2e} - \frac{m_{11}}{m_{22}}u(z_6 + \alpha_3)\right.$$
$$+ \mu S\dot{z}_{2e} + \frac{P(t)}{m_{22}}\tau_v + \mu\dot{S}z_{2e} - \dot{\alpha}_2\}.$$

Since $\mu S\dot{z}_{2e} + \mu\dot{S}z_{2e} - \dot{\alpha}_2 = \mu\dot{z}_{2e}$, we have

$$\dot{V}_1 = -\varepsilon_1\lambda_1 z_{1e}^2 - \mu\lambda_1 z_{2e}^2 + \lambda_1 z_{2e}z_5 + \alpha_3(\lambda_1 z_{1e}z_2 - \lambda_1 z_1 z_{2e}$$
$$- \frac{\lambda_2 m_{11}}{m_{22}}uz_5 - \lambda_2\mu z_1 z_5) + z_4\left[\frac{m_{22}}{m_{11}}vr - \frac{d_{11}}{m_{11}}u + \frac{1}{m_{11}}\tau_u - \dot{u}_d + \lambda_1 z_{1e}\right.$$
$$- \lambda_2(1 - S(t))\frac{m_{11}}{m_{22}}r_d z_5 - \dot{\alpha}_1 - \lambda_2 Q(t)S(t)\frac{m_{11}}{m_{22}}z_5 r_d]$$
$$+ \lambda_2 z_5\left[-P(t)S(t)\frac{m_{11}}{m_{22}}z_4 r_d - \frac{m_{11}}{m_{22}}\alpha_1 r_d - \frac{d_{22}}{m_{22}}(z_5 + \alpha_2)\right.$$
$$+ \frac{\mu S d_{22}}{m_{22}}z_{2e} - P(t)S(t)\frac{m_{11}}{m_{22}}uz_6 + \mu(z_5 + \alpha_2)$$
$$- \mu^2 S z_{2e} - \mu z_{1e}r_d + \frac{P(t)}{m_{22}}\tau_v]$$
$$+ z_6[\lambda_1 z_{1e}z_2 - \lambda_1 z_1 z_{2e} - \mu\lambda_2 z_1 z_5$$
$$- (1 - S(t))\frac{m_{11}}{m_{22}}\lambda_2 uz_5 - \lambda_2 Q(t)S(t)\frac{m_{11}}{m_{22}}uz_5].$$

Note that $P^2(t) = P(t)$, and we choose torques for τ_u and τ_v as

$$\begin{cases} \tau_u = m_{11}[-\varepsilon_2 z_4 - \dfrac{m_{22}}{m_{11}}vr + \dfrac{d_{11}}{m_{11}}u - \lambda_1 z_{1e} + \lambda_2(1 - S(t))\dfrac{m_{11}}{m_{22}}r_d z_5 \\ \quad + \lambda_2 Q(t)S(t)\dfrac{m_{11}}{m_{22}}r_d z_5 + \dot{\alpha}_1 + \dot{u}_d] \\ \tau_v = P(t)m_{22}[-\varepsilon_3 z_5 + S(t)\dfrac{m_{11}}{m_{22}}z_4 r_d + S(t)\dfrac{m_{11}}{m_{22}}uz_6] \end{cases}$$

$$(9.14)$$

where ε_2 and ε_3 are positive constants. We get

$$\dot{V}_1 = -\varepsilon_1\lambda_1 z_{1e}^2 - \mu\lambda_1 z_{2e}^2 + \lambda_1 z_{2e}z_5 + \alpha_3(\lambda_1 z_{1e}z_2$$
$$- \lambda_1 z_1 z_{2e} - \frac{\lambda_2 m_{11}}{m_{22}}uz_5 - \lambda_2\mu z_1 z_5)$$
$$- \varepsilon_2 z_4^2 - (P(t)\varepsilon_3 + \lambda_2\frac{d_{22}}{m_{22}} - \lambda_2\mu)z_5^2 + \lambda_2 z_5[-\frac{m_{11}}{m_{22}}\alpha_1 r_d + \frac{\mu d_{22}}{m_{22}}z_{2e}$$
$$- \mu^2 z_{2e} - \mu z_{1e}r_d] + z_6[\lambda_1 z_{1e}z_2 - \lambda_1 z_1 z_{2e} - \mu\lambda_2 z_1 z_5$$
$$- (1 - S(t))\frac{m_{11}}{m_{22}}\lambda_2 uz_5 - \lambda_2 Q(t)S(t)\frac{m_{11}}{m_{22}}uz_5].$$

Define a new Lyapunov function $V_2 = V_1 + \dfrac{1}{2}z_{3e}^2$ and choose the virtual controller α_3 as

$$\alpha_3 = -\varepsilon_3\left(z_{3e} + \lambda_1 z_{1e}z_2 - \lambda_1 z_1 z_{2e} - \frac{\lambda_2 m_{11}}{m_{22}}uz_5 - \lambda_2\mu z_1 z_5\right) \tag{9.15}$$

where ε_3 is a positive constant, then we get

$$
\begin{aligned}
\dot{V}_2 =\ & -\varepsilon_1\lambda_1 z_{1e}{}^2 - \mu\lambda_1 z_{2e}{}^2 \\
& + \lambda_1 z_{2e}z_5 - \varepsilon_3\left(z_{3e} + \lambda_1 z_{1e}z_2 - \lambda_1 z_1 z_{2e} - \frac{\lambda_2 m_{11}}{m_{22}}uz_5 - \lambda_2\mu z_1 z_5\right)^2 \\
& - \varepsilon_2 z_4^2 - \left(P(t)\varepsilon_3 + \lambda_2\frac{d_{22}}{m_{22}} - \lambda_2\mu\right)z_5^2 \\
& + \lambda_2 z_5\left[-\frac{m_{11}}{m_{22}}\alpha_1 r_d + \frac{\mu d_{22}}{m_{22}}z_{2e} - \mu^2 z_{2e} - \mu z_{1e}r_d\right] \\
& + z_6\left[\lambda_1 z_{1e}z_2 - \lambda_1 z_1 z_{2e} - \mu\lambda_2 z_1 z_5\right. \\
& \left. - (1 - S(t))\frac{m_{11}}{m_{22}}\lambda_2 uz_5 - \lambda_2 Q(t)S(t)\frac{m_{11}}{m_{22}}uz_5\right].
\end{aligned}
$$

Now consider the following Lyapunov candidate for the closed-loop system $V_3 = V_2 + \dfrac{1}{2}z_6^2$ and choose the yaw torque as

$$
\begin{aligned}
\tau_r = m_{33}\Big[& -\varepsilon_4 z_6 - \frac{m_{11} - m_{22}}{m_{33}}uv + \frac{d_{33}}{m_{33}}r_e - \lambda_1 z_{1e}z_2 + \lambda_1 z_1 z_{2e} \\
& + \mu\lambda_2 z_1 z_5 + (1 - S(t))\frac{m_{11}}{m_{22}}\lambda_2 uz_5 + \lambda_2 Q(t)S(t)\frac{m_{11}}{m_{22}}uz_5 + \dot{\alpha}_3 + \dot{r}_d\Big] \tag{9.16}
\end{aligned}
$$

where ε_4 is a positive constant, then we get

$$
\begin{aligned}
\dot{V}_3 =\ & -\varepsilon_1\lambda_1 z_{1e}{}^2 - \mu\lambda_1 z_{2e}{}^2 \\
& + \lambda_1 z_{2e}z_5 - \varepsilon_3\left(z_{3e} + \lambda_1 z_{1e}z_2 - \lambda_1 z_1 z_{2e} - \frac{\lambda_2 m_{11}}{m_{22}}uz_5 - \lambda_2\mu z_1 z_5\right)^2 \\
& - \varepsilon_2 z_4^2 - \left(p\varepsilon_3 + \lambda_2\frac{d_{22}}{m_{22}} - \lambda_2\mu\right)z_5^2 + \lambda_2 z_5\left[-\frac{m_{11}}{m_{22}}\alpha_1 r_d + \frac{\mu d_{22}}{m_{22}}z_{2e}\right. \\
& \left. - \mu^2 z_{2e} - \mu z_{1e}r_d\right] - \varepsilon_4 z_6^2.
\end{aligned}
$$

Define $r_d{}^{\max}$ as the maximum value of $|r_d(t)|$, and we get

$$
\begin{aligned}
& \lambda_1 z_{2e}z_5 + \lambda_2 z_5\left[-\frac{m_{11}}{m_{22}}\alpha_1 r_d + \frac{\mu d_{22}}{m_{22}}z_{2e} - \mu^2 z_{2e} - \mu z_{1e}r_d\right] \\
& \leq \frac{\lambda_2\varepsilon_1}{2}|r_d|\left(\frac{\varepsilon_1 m_{11}}{m_{22}} - \mu\right)\left(\vartheta_1 z_5^2 + \frac{z_{1e}{}^2}{\vartheta_1}\right) \\
& + \frac{\lambda_2}{2}\left(\frac{\mu d_{22}}{m_{22}} - \mu^2\right)\left(\vartheta_2 z_5^2 + \frac{z_{2e}{}^2}{\vartheta_2}\right) + \frac{\lambda_1}{2}\left(\vartheta_3 z_5^2 + \frac{z_{2e}{}^2}{\vartheta_3}\right)
\end{aligned}
$$

if $\mu < \dfrac{d_{22}}{m_{22}}$ and $\mu < \dfrac{\varepsilon_1 m_{11}}{m_{22}}$, where ϑ_i, $1 \leq i \leq 3$ are positive constants.
If we choose ε_1, λ_1, λ_2, μ, $r_d{}^{\max}$ and ϑ_i, $1 \leq i \leq 3$ such that

$$
\begin{cases}
\varepsilon_1 \lambda_1 - \dfrac{\lambda_2 \varepsilon_1 m_{11} r_d{}^{\max}}{2\vartheta_1 m_{22}} + \dfrac{\lambda_2 \mu r_d{}^{\max}}{2\vartheta_1} := k_1 > 0 \\[2mm]
\lambda_1 \mu - \dfrac{\lambda_2 \mu d_{22}}{2\vartheta_2 m_{22}} + \dfrac{\lambda_2 \mu^2}{2\vartheta_2} - \dfrac{\lambda_1}{2\vartheta_3} := k_2 > 0 \\[2mm]
\dfrac{\lambda_2 d_{22}}{m_{22}} + p\varepsilon_3 - \mu\lambda_2 \\[2mm]
- \dfrac{\lambda_2 \varepsilon_1 \vartheta_1 m_{11}}{2m_{22}} r_d{}^{\max} - \dfrac{\lambda_2 \mu \vartheta_2 d_{22}}{2m_{22}} + \dfrac{\lambda_2 \vartheta_2 \mu^2}{2} + \dfrac{\lambda_2 \mu r_d{}^{\max} \vartheta_1}{2} \\[2mm]
- \dfrac{\vartheta_3 \lambda_1}{2} := k_3 > 0
\end{cases}
\tag{9.17}
$$

then we get

$$
\dot{V}_4 \leq - k_1 z_{1e}{}^2 - k_2 z_{2e}{}^2 - \varepsilon_3(z_{3e} + \lambda_1 z_{1e} z_2
$$
$$
- \lambda_1 z_1 z_{2e} - \dfrac{\lambda_2 m_{11}}{m_{22}} u z_5 - \lambda_2 \mu z_1 z_5)^2
$$
$$
- \varepsilon_2 z_4{}^2 - k_3 z_5{}^2 - \varepsilon_4 z_6{}^2.
\tag{9.18}
$$

It could be checked that this condition can always be fulfilled if we choose μ as being close enough but small relative to $\frac{d_{22}}{m_{22}}$; r_d is carefully designed such that $r_d{}^{\max}$ is small enough; ε_1 is free to design. When the surface vessel is fully-actuated, then there is no restrictive condition on r_d.

Theorem 9.1. *If there is a $r_d{}^{\max}$ such that (9.17) holds, then global uniform tracking control for a fully actuated and underactuated surface vessel is solved by controllers (9.14) and (9.16) under assumption 9.1. In particular, all tracking errors converge to the origin as $t \to \infty$.*

Proof:
Since $\dot{V}_3(t) \leq 0$, thus $z_{1e}, z_{2e}, \alpha_3, z_4, z_4, z_6$ are bounded. Also $\dot{V}_3(t)$ is bounded on $[0, \infty)$ and $V_3(t)$ is continuous. In addition, (9.18) yields that $V_3(t)$ is integrable on $[0, \infty)$. With Barbalat's lemma we know $V_3(t) \to 0$ as $t \to \infty$. Finally $\alpha_3 \to 0$ yields $z_{3e} \to 0$.

9.4 Simulation Results

In this section, we carry out a simulation to illustrate the effectiveness and performance of the proposed controller. We use a circle as a reference trajectory which has been considered in many papers. We make the choice of initial conditions for reference trajectory as

$$
(x_d(0), y_d(0), \phi_d(0), v_d(0)) = (0, 0, 0, 0.1).
$$

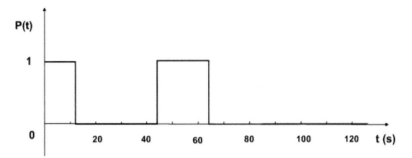

Fig. 9.3. A schematic figure of P(t).

The reference torques are chosen as $u_d = 0.1$, $r_d = 0.1$. The initial condition of the vessel system is

$$(x(0), y(0), \phi(0), u(0), v(0), r(0)) = (1, -1, 0.5, 0, 0, 0).$$

The parameters are chosen as $\mu = 0.3$, $\lambda_1 = 0.5$, $\lambda_2 = 0.2$. The parameters of the vessel are listed as: $d_{11} = d_{33} = 0$, $d_{22} = 0.4$, $m_{11} = m_{22} = m_{33} = 0.1$. β is chosen to be 2. It is easy to check that (9.17) is fulfilled. $P(t)$ is shown in Fig.9.3. It is shown that the proposed method yields a satisfactory tracking result. Fig.9.4 shows the position of the vessel. Fig.9.5 shows the trajectory tracking errors. Fig.9.6 shows the torque of the vessel. Fig.9.7 shows the velocities u, v and r of the vessel.

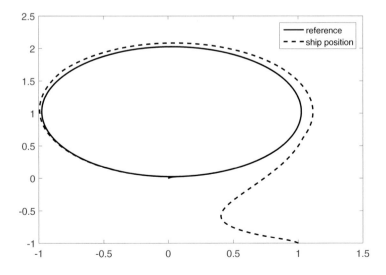

Fig. 9.4. Simulation result of vessel position.

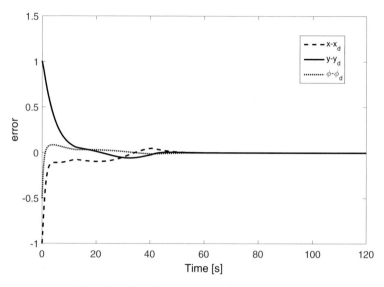

Fig. 9.5. Simulation result of tracking errors.

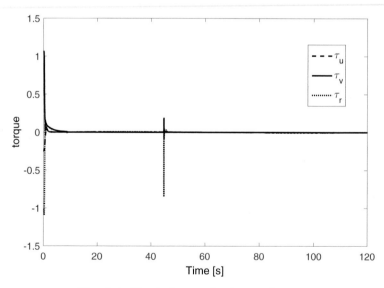

Fig. 9.6. Simulation result of control torques.

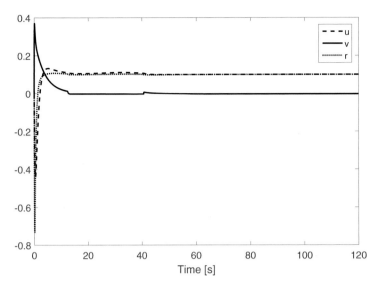

Fig. 9.7. The velocities u, v and r of the vessel.

9.5 Conclusion

In this chapter we investigate the fault-tolerant tracking control for an underactuated surface vessel without actuator redundancy. As illustrated by the simulation, the control scheme can deal with both cases when sway force suffers from actuator failure or is turned on and off on purpose.

Part IV

Adaptive Control of Multiple Underactuated Nonlinear Systems

10

Adaptive Formation Control of Multiple Nonholonomic Mobile Robots

In this chapter, we investigate the formation problem of tracking a desired trajectory for a group of nonholonomic mobile robots with unknown parameters. The mobile robot systems are allowed to have non-identical dynamics where only a subset of the mobile robot systems can obtain the desired trajectory information directly. In spite of this, a distributed adaptive control approach based on backstepping technique is proposed with which local controllers are designed by utilizing only the information collected within neighboring areas. By introducing the estimates to account for the parametric uncertainties of its neighbors' dynamics into the local controller of each subsystem, the transmission of parameter estimates among linked subsystems can be avoided. It is proved that the boundedness of all closed-loop signals and the asymptotically consensus tracking for all the robots' outputs are ensured.

10.1 Introduction

Because of its widespread potential applications in various fields such as mobile robot networks, intelligent transportation management, surveillance and monitoring, distributed coordination of multiple dynamic subsystems (also known as multi-agent systems) has achieved rapid development during the past decades. Formation control is one of the most popular topics in this area, which has received significant attention by numerous researchers. It is often aimed to achieve an agreement for certain variables, such as the states or the outputs, of the subsystems in a group. A large number of effective control approaches have been proposed to solve the constant consensus problems and the non-constant consensus issues of tracking time-varying trajectories. According to whether the desired consensus values are determined by exogenous inputs, which are sometimes regarded as virtual leaders, these approaches are also classified as leaderless consensus and leader-following consensus solutions.

Note that the considered robots are actually uncertain underactuated mechanical systems with both dynamic and kinematic models, which brings new difficulties in designing distributed adaptive controllers. Therefore, only a few results have been reported in this area so far [11, 12, 13]. In [11], the formation control of multiple unicycle-type mobile robots at the dynamic model level is investigated. A path-following approach by combining the virtual structure technique is presented to derive the formation architecture. In [12], a formation control scheme is proposed for multiple mobile robots and no collision between any two robots is guaranteed. In the two schemes, all the robots require the exact information of the reference trajectory. In [13], the flocking control of a collection of nonholonomic mobile robots is proposed, where only part of the robots can obtain the exact knowledge of the reference directly. However, the system model considered is limited at the kinematic level. Motivated by these, we investigate the formation control problem for multiple nonholonomic mobile robots at dynamic model level with unknown parameters under the assumption that only part of the robots can access the exact information of the reference directly. Such a challenging problem can be regarded as a generalized problem of one-dimensional output consensus tracking by considering demanding distances in the X-Y plane. It is shown that the problem can be successfully solved with the combination of our proposed distributed control strategy and the transverse function technique in [15]. The formation errors of the overall systems can be made as small as desired by adjusting the design parameters properly.

10.2 Problem Formulation

Suppose that the communications among the N subsystems can be represented by a directed graph $\mathcal{G} \leq (\mathcal{V}, \mathcal{E})$ where $\mathcal{V} = \{1, \ldots, N\}$ denotes the set of indexes (or vertices) corresponding to each subsystem, $\mathcal{E} \subseteq \mathcal{V} \times \mathcal{V}$ is the set of edges between two distinct subsystems. An edge $(i, j) \in \mathcal{E}$ indicates that subsystem j can obtain information from subsystem i, but not necessarily vice versa [72]. In this case, subsystem i is called a neighbor of subsystem j. We denote the set of neighbors for subsystem i as $\mathcal{N}_i$. Self edges (i, i) is not allowed, thus $(i, i) \notin \mathcal{E}$ and $i \notin \mathcal{N}_i$. The connectivity matrix $A = [a_{ij}] \in \Re^{N \times N}$ is defined such that $a_{ij} = 1$ if $(j, i) \in \mathcal{E}$ and $a_{ij} = 0$ if $(j, i) \notin \mathcal{E}$. Clearly, the diagonal elements $a_{ii} = 0$. We introduce an in-degree matrix $\triangle$ such that $\triangle = \mathrm{diag}(\triangle_i) \in \Re^{N \times N}$ with $\triangle_i = \sum_{j \in \mathcal{N}_i} a_{ij}$ being the ith row sum of A. Then, the Laplacian matrix of $\mathcal{G}$ is defined as $\mathcal{L} = \triangle - A$.

We now use $\mu_i = 1$ to indicate the case that x_r and y_r are accessible directly to each robot i; otherwise, μ_i is set as $\mu_i = 0$. Based on this, the *control objective* is to design distributed adaptive controllers u_i for each robot by utilizing only locally available information obtained from the intrinsic robot and its neighbors such that:

(i) all the signals in the closed-loop system are globally uniformly bounded;

(ii) the outputs of all robots can still track the desired trajectory $x_r(t)$ and $y_r(t)$ asymptotically, though $\mu_i = 1$ only for some $i \in \{1, 2, \ldots, N\}$.

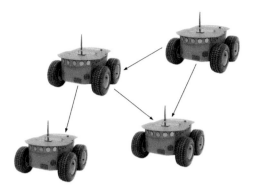

Fig. 10.1. A schematic diagram of multiple mobile robots.

To achieve the objective, the following assumptions are imposed.

Assumption 10.1 *The first nth-order derivatives of $f_r(t)$ are bounded, piecewise continuous and known to all subsystems in the group.*

Assumption 10.2 *The sign of b_i is available in constructing u_i for subsystem i and $\beta_i(x_i) \neq 0$.*

In this section, the proposed distributed adaptive tracking control strategy will be applied to solve a formation control problem for multiple nonholonomic mobile robots at dynamic model level with unknown parameters. In contrast to the existing results such as [11, 12, 13], the formation control problem for a group of nonholonomic mobile robots at dynamic model level with unknown parameters will be investigated under the assumption that only part of the robots can access the exact information of the reference directly. Another key technique for solving this problem is the transverse function approach [15] with which the underactuated system can be transformed to a fully actuated one. Based on this, the control objectives for the position (including $\bar{x}_i$ and $\bar{y}_i$, i.e. the components in X and Y directions, respectively) and the orientation ($\bar{\phi}_i$) of the ith robot can be treated separately.

We consider a group of N two-wheeled mobile robots as schematically shown in Fig. 10.1. For the ith robot, point P_{ic} represents the center of the mass, P_{io} is the middle point located on the straight line connecting the left and right wheels and ι_i denotes the distance between P_{ic} and P_{io}. b_i is the half width of the robot, r_i is the radius of the wheel. OXY is the earth-fixed coordinate system. $(\bar{x}_i, \bar{y}_i, \bar{\phi}_i)$ denotes the position and orientation of the ith robot. According to [11], the ith mobile robot can be described by the following dynamic model

$$\dot{\eta}_i = J(\eta_i)\omega_i \tag{10.1}$$

$$M_i\dot{\omega}_i + C_i(\dot{\eta}_i)\omega_i + D_i\omega_i = \tau_i, \quad \text{for } i = 1, \dots, N \tag{10.2}$$

where $\eta_i = [\bar{x}_i, \bar{y}_i, \bar{\phi}_i]^T$, $\omega_i = [\omega_{i1}, \omega_{i2}]^T$ denotes the angular velocities of the left and right wheels, $\tau_i = [\tau_{i1}, \tau_{i2}]^T$ represents the control torques applied to the wheels. M_i is a symmetric, positive definite inertia matrix, $C_i(\dot{\eta}_i)$ is the centripetal and coriolis matrix, D_i denotes the surface friction. These matrices have the same form as those in [11], which are given below for completeness.

$$J(\eta_i) = \frac{r_i}{2}\begin{bmatrix} \cos\bar{\phi}_i & \cos\bar{\phi}_i \\ \sin\bar{\phi}_i & \sin\bar{\phi}_i \\ b_i^{-1} & -b_i^{-1} \end{bmatrix}, \quad M_i = \begin{bmatrix} m_{i1} & m_{i2} \\ m_{i2} & m_{i1} \end{bmatrix}$$

$$C_i(\dot{\eta}_i) = \begin{bmatrix} 0 & c_i\dot{\bar{\phi}}_i \\ -c_i\dot{\bar{\phi}}_i & 0 \end{bmatrix}, \quad D_i = \begin{bmatrix} d_{i1} & 0 \\ 0 & d_{i2} \end{bmatrix}$$

$$m_{i1} = \frac{1}{4}b_i^{-2}r_i^2(m_i b_i^2 + I_i) + I_{wi}$$

$$m_{i2} = \frac{1}{4}b_i^{-2}r_i^2(m_i b_i^2 - I_i)$$

$$I_i = m_{ci}l_i^2 + 2m_{wi}b_i^2 + I_{ci} + 2I_{mi}$$

$$c_i = \frac{1}{2}b_i^{-1}r_i^2 m_{ci}l_i, \quad m_i = m_{ci} + 2m_{wi}. \tag{10.3}$$

In (10.3), m_{ci}, m_{wi}, I_{ci}, I_{wi}, I_{mi} and d_{ik} are unknown system parameters of which the physical meanings can be found in [11].

Remark 10.1. Observing (10.1) with $J(\eta_i)$ in (10.3), it can be seen that the number of inputs (i.e. ω_{i1} and ω_{i2}) is less than the number of configurations (i.e. $\bar{x}_i$, $\bar{y}_i$ and $\bar{\phi}_i$). Thus the considered mobile robot is an underactuated mechanical system. To achieve the tracking objectives of $\bar{x}_i$, $\bar{y}_i$ and $\bar{\phi}_i$ separately, a transverse function approach [15] will be employed. An auxiliary manipulated variable will be introduced with which the underactuated problem can be transformed to a fully-actuated one.

10.2.1 Change of Coordinates

We change the original coordinates of the ith robot as follows:

$$\begin{bmatrix} x_i \\ y_i \end{bmatrix} = \begin{bmatrix} \bar{x}_i \\ \bar{y}_i \end{bmatrix} + R(\phi_i)\begin{bmatrix} f_{1i}(\xi_i) \\ f_{2i}(\xi_i) \end{bmatrix} \tag{10.4}$$

$$\phi_i = \bar{\phi}_i - f_{3i}(\xi_i) \tag{10.5}$$

where

$$R(\phi_i) = \begin{bmatrix} \cos(\phi_i) & -\sin(\phi_i) \\ \sin(\phi_i) & \cos(\phi_i) \end{bmatrix} \tag{10.6}$$

and $f_{li}(\xi_i)$ for $l = 1, 2, 3$ are functions of ξ_i designed as

$$f_{1i}(\xi_i) = \varepsilon_{1i}\sin(\xi_i)\frac{\sin(f_{3i})}{f_{3i}}$$

$$f_{2i}(\xi_i) = \varepsilon_{1i}\sin(\xi_i)\frac{1-\cos(f_{3i})}{f_{3i}}$$

$$f_{3i}(\xi_i) = \varepsilon_{2i}\cos(\xi_i) \tag{10.7}$$

where ε_{1i} and ε_{2i} are positive constants and ε_{2i} satisfies $0 < \varepsilon_{2i} < \frac{\pi}{2}$. It can be shown that

$$|f_{1i}| < \varepsilon_{1i}, |f_{2i}| < \varepsilon_{1i}, |f_{3i}| < \varepsilon_{2i}. \tag{10.8}$$

Computing the derivatives of x_i, y_i and ϕ_i yields that

$$\begin{bmatrix} \dot{x}_i \\ \dot{y}_i \end{bmatrix} = Q_i \begin{bmatrix} r_i u_{i1} \\ \dot{\xi}_i \end{bmatrix} + \frac{\partial R(\phi_i)}{\partial \phi_i} \begin{bmatrix} f_{1i}(\xi_i) \\ f_{2i}(\xi_i) \end{bmatrix}$$

$$\times \left(r_i b_i^{-1} u_{i2} - \frac{\partial f_{3i}(\xi_i)}{\partial \xi_i}\dot{\xi}_i \right) \tag{10.9}$$

$$\dot{\phi}_i = r_i b_i^{-1} u_{i2} - \frac{\partial f_{3i}(\xi_i)}{\partial \xi_i}\dot{\xi}_i \tag{10.10}$$

where $u_{i1} = 0.5(\omega_{i1} + \omega_{i2})$ and $u_{i2} = 0.5(\omega_{i1} - \omega_{i2})$.

$$Q_i = \left[\begin{pmatrix} \cos(\bar{\phi}_i) \\ \sin(\bar{\phi}_i) \end{pmatrix} \quad R(\phi_i) \begin{pmatrix} \dfrac{\partial f_{1i}(\xi_i)}{\partial \xi_i} \\ \dfrac{\partial f_{2i}(\xi_i)}{\partial \xi_i} \end{pmatrix} \right] \tag{10.11}$$

is ensured to be invertible [15]. Different from $(\bar{x}_i, \bar{y}_i, \bar{\phi}_i)$, the transformed coordinates $(x_i, y_i$ and $\phi_i)$ can be controlled separately by tuning u_{i1}, u_{i2} and $\dot{\xi}_i$ which is deemed as an auxiliary manipulated variable.

Remark 10.2. With the transverse functions (10.7), u_{i1}, $\dot{\xi}_i$ and u_{i2} act as the control input for the MIMO kinematic model (10.9) and (10.10) of the mobile robot while x_i, y_i and ϕ_i are the output. Thus does not like the traditional underactuated kinematic mobile robot model, the MIMO can be treated as three separate SISO systems.

10.2.2 Formation Control Objective

The components of the desired trajectory in X and Y directions can be expressed as

$$x_r(t) = w_r f_{rx}(t) + c_{rx} \quad y_r(t) = w_r f_{ry}(t) + c_{ry}. \tag{10.12}$$

It is assumed that $f_{rx}(t)$ and $f_{ry}(t)$ are known by all the robots, whereas the parameters w_r, c_{rx} and c_{ry} are only available to part of the robots. Besides, $\phi_r(t) \le \arctan\left(\frac{\dot{y}_r}{\dot{x}_r}\right)$ denotes the reference trajectory for the orientation of each robot.

The *control objective* in this section is to design distributed adaptive formation controllers such that all the robots can follow a desired trajectory in X-Y plane by maintaining certain prescribed demanding distances from the desired trajectory, i.e.

$$\lim_{t \to \infty} x_i(t) - x_r(t) = -\rho_{ix} \tag{10.13}$$

$$\lim_{t \to \infty} y_i(t) - y_r(t) = -\rho_{iy} \tag{10.14}$$

$$\lim_{t \to \infty} \phi_i(t) - \phi_r(t) = 0. \tag{10.15}$$

Remark 10.3. It can be seen that the consensus tracking objective (ii), i.e. $\lim_{t \to \infty} [y_i(t) - y_r(t)] = 0$, is actually a special case of the formation objectives in (10.13) or (10.14) with $\rho_{ix} = 0$ or $\rho_{iy} = 0$. Note that in contrast to the fact that exact information about $x_r(t)$ and $y_r(t)$ are only accessible to a subset of the robots, the desired orientation $\phi_r(t) = \arctan\left(\frac{\dot{y}_r}{\dot{x}_r}\right)$ is available to all the robots since $f_{rx}(t)$ and $f_{ry}(t)$ are available to all the robots.

Moreover, from (10.4), (10.5) and the properties of f_{li} in (10.8), it is clear that the transformation errors $x_i - \bar{x}_i$, $y_i - \bar{y}_i$, $\phi_i - \bar{\phi}_i$ are bounded by ε_{1i} and ε_{2i}. It will be shown that the designed distributed adaptive controllers can guarantee the convergence of the formation control errors with respect to x_i, y_i and ϕ_i. Therefore, the formation control errors with respect to the true position and orientation, i.e. $\bar{x}_i$, $\bar{y}_i$ and $\bar{\phi}_i$, can be made as small as desired by adjusting ε_{1i} and ε_{2i} properly.

We suppose that the communication status among the N robots can be represented by a directed graph $\mathcal{G}$ and Assumption 10.2 holds. To achieve the formation control objective, the following assumptions are also needed.

Assumption 10.3 f_{rx}, f_{ry}, $\dot{f}_{rx}$, $\dot{f}_{ry}$ and $\ddot{f}_{rx}$, $\ddot{f}_{ry}$ *are bounded, piece-wise continuous bounded and known to all the robots.*

Assumption 10.4 *The parameters r_i and b_i fall in known compact sets, i.e. there exist some known positive constants $\bar{r}_i$, $\underline{r}_i$, $\bar{b}_i$ and $\underline{b}_i$ such that $\underline{r}_i < r_i < \bar{r}_i$ and $\underline{b}_i < b_i < \bar{b}_i$.*

Assumption 10.5 *The demanding distances ρ_{ix} and ρ_{iy} for robot i are available to its neighbors.*

The following lemma brought from [97] is then introduced, which will be useful in our design and analysis of the distributed adaptive controllers.

Lemma 10.4. [97] *Based on Assumption 3, the matrix $(\mathcal{L} + \mathcal{B})$ is nonsingular where $\mathcal{B} = diag\{\mu_1, \ldots, \mu_N\}$. Define*

$$\bar{q} = [\bar{q}_1, \ldots, \bar{q}_N]^T = (\mathcal{L} + \mathcal{B})^{-1}[1, \ldots, 1]^T$$

$$P = diag\{P_1, \ldots, P_N\} = diag\left\{\frac{1}{\bar{q}_1}, \ldots, \frac{1}{\bar{q}_N}\right\}$$

$$Q = P(\mathcal{L} + \mathcal{B}) + (\mathcal{L} + \mathcal{B})^T P, \tag{10.16}$$

then $\bar{q}_i > 0$ for $i = 1, \ldots, N$ and Q is positive definite.

10.3 Control Design

Observed from (10.1), (10.2), (10.4)-(10.10), the relative degree from the actual control inputs τ_i to the output (x_i, y_i, ϕ_i) is two. Thus the control design procedure involves two steps by adopting the backstepping technique. In the first step, the virtual controls for u_{i1}, u_{i2} and the auxiliary manipulated variable $\dot{\xi}_i$ will be chosen. In the second step, the actual control inputs τ_i will be derived.

• *Step 1*: Define local error variables as

$$z_{ix,1} = \sum_{j=1}^{N} a_{ij}(x_i + \rho_{ix} - x_j - \rho_{jx}) + \mu_i(x_i + \rho_{ix} - x_r)$$

$$z_{iy,1} = \sum_{j=1}^{N} a_{ij}(y_i + \rho_{iy} - y_j - \rho_{jy}) + \mu_i(y_i + \rho_{iy} - y_r)$$

$$e_{ix,1} = x_i - \mu_i x_r - (1 - \mu_i)(f_{rx}\hat{w}_{rx,i} - \hat{c}_{rx,i}) + \rho_{ix}$$

$$e_{iy,1} = y_i - \mu_i y_r - (1 - \mu_i)(f_{ry}\hat{w}_{ry,i} - \hat{c}_{ry,i}) + \rho_{iy}$$

$$\delta_{i\phi} = \phi_i - \phi_r$$

$$e_{ix,2} = u_{i1} - \alpha_{i1}, \ e_{i\phi,2} = u_{i2} - \alpha_{i2} \qquad (10.17)$$

where $\hat{w}_{rx,i}$ ($\hat{w}_{ry,i}$), $\hat{c}_{rx,i}$ and $\hat{c}_{ry,i}$ are the estimates introduced in the ith robot for the unknown trajectory parameters if $\mu_i = 0$. We choose the virtual controls $(\alpha_{i1}, \alpha_{i2})$ and $\dot{\xi}$ in transverse function technique as

$$\begin{bmatrix} \alpha_{i1} \\ \dot{\xi}_i \end{bmatrix} = \begin{bmatrix} \hat{\theta}_{i1}^{-1} & 0 \\ 0 & 1 \end{bmatrix} Q_i^{-1} \Omega_i \qquad (10.18)$$

$$\alpha_{i2} = \hat{\theta}_{i2}^{-1}\left(-k_2\delta_{i\phi} + \frac{\partial f_{3i}(\xi_i)}{\partial \xi_i}\dot{\xi}_i + \dot{\phi}_r\right) \qquad (10.19)$$

where $\hat{\theta}_{i1}$ and $\hat{\theta}_{i2}$ are the estimates of r_i and $r_i b_i^{-1}$, respectively.

$$\Omega_i = -k_1 P_i \begin{bmatrix} z_{ix,1} \\ z_{iy,1} \end{bmatrix} - \frac{\partial R(\phi_i)}{\partial \phi_i}\begin{bmatrix} f_{1i}(\xi_i) \\ f_{2i}(\xi_i) \end{bmatrix}\left(-k_2\delta_{i\phi}\right.$$

$$+\dot{\phi}_r\left.\right) - \begin{bmatrix} \dot{\rho}_{ix} \\ \dot{\rho}_{iy} \end{bmatrix} + \mu_i\begin{bmatrix} \dot{f}_{rx}w_{rx} \\ \dot{f}_{ry}w_{ry} \end{bmatrix}$$

$$+(1 - \mu_i)\begin{bmatrix} \dot{f}_{rx}\hat{w}_{rx,i} + f_{rx}\dot{\hat{w}}_{rx,i} + \dot{\hat{c}}_{rx,i} \\ \dot{f}_{ry}\hat{w}_{ry,i} + f_{ry}\dot{\hat{w}}_{ry,i} + \dot{\hat{c}}_{ry,i} \end{bmatrix}$$

where k_1, k_2 are positive constants. P_i is defined in (10.16). The above design delivers the following results.

$$\begin{bmatrix} \dot{e}_{ix,1} \\ \dot{e}_{iy,1} \end{bmatrix} = -k_1 P_i \begin{bmatrix} z_{ix,1} \\ z_{iy,1} \end{bmatrix} + \frac{\partial R(\phi_i)}{\partial \phi_i} \begin{bmatrix} f_{1i}(\xi_i) \\ f_{2i}(\xi_i) \end{bmatrix} \left(\tilde{\theta}_{i2} u_{i2} \right.$$

$$\left. + \hat{\theta}_{i2} e_{i\phi,2} \right) + Q_i \begin{bmatrix} \tilde{\theta}_{i1} u_{i1} + \hat{\theta}_{i1} e_{ix,2} \\ 0 \end{bmatrix}$$

$$\dot{\delta}_{i\phi} = -k_2 \delta_{i\phi} + \tilde{\theta}_{i2} u_{i2} + \hat{\theta}_{i2} e_{i\phi,2} \tag{10.20}$$

The parameter estimators at this step are designed as

$$\dot{\hat{w}}_{rx,i} = -\gamma_{ri} f_{rx} e_{ix,1}, \quad \dot{\hat{w}}_{ry,i} = -\gamma_{ri} f_{ry} e_{iy,1}$$

$$\dot{\hat{c}}_{rx,i} = -\gamma_{ri} e_{ix,1}, \quad \dot{\hat{c}}_{ry,i} = -\gamma_{ri} e_{iy,1}$$

$$\dot{\hat{\theta}}_{i1} = \text{Proj}\left(\hat{\theta}_{i1}, \gamma_{\theta_{i1}} \pi_{i1} \right), \quad \dot{\hat{\theta}}_{i2} = \text{Proj}\left(\hat{\theta}_{i2}, \gamma_{\theta_{i1}} \pi_{i2} \right)$$

$$\pi_{i1} = \left[e_{ix,1} \cos(\bar{\phi}_i) + e_{iy,1} \sin(\bar{\phi}_i) \right] u_{i1}$$

$$\pi_{i2} = [e_{ix,1}, e_{iy,1}] \frac{\partial R(\phi_i)}{\partial \phi_i} \begin{bmatrix} f_{1i} \\ f_{2i} \end{bmatrix} u_{i2} + \delta_{i\phi} u_{i2} \tag{10.21}$$

Note $\text{Proj}(\cdot, \cdot)$ denotes a Lipschitz continuous projection operator about which the design details can be found in [93]. It is adopted here to ensure that $\hat{\theta}_{i1} > 0$ and $\hat{\theta}_{i2} > 0$. Thus $\hat{\theta}_{i1}^{-1}$ and $\hat{\theta}_{i2}^{-1}$ in (10.18) and (10.19) are well defined.

We choose a Lyapunov function candidate at this step as

$$V_1 = \frac{1}{2} \sum_{i=1}^{N} \left(e_{ix,1}^2 + e_{iy,1}^2 + \delta_{i\phi}^2 + \frac{1}{2\gamma_{\theta_{i1}}} \tilde{\theta}_{i1}^2 + \frac{1}{2\gamma_{\theta_{i2}}} \tilde{\theta}_{i2}^2 \right)$$

$$+ \frac{k_1}{2} \sum_{i=1}^{N} (1 - \mu_i) \frac{P_i}{\gamma_{ri}} \left(\tilde{w}_{rx,i}^2 + \tilde{w}_{ry,i}^2 + \tilde{c}_{rx,i}^2 + \tilde{c}_{ry,i}^2 \right). \tag{10.22}$$

From (10.20) and (10.21), the derivative of V_1 in (10.22) can be computed as

$$\dot{V}_1 \leq -\frac{k_1}{2} \left(\delta_x^T Q \delta_x + \delta_y^T Q \delta_y \right) - k_2 \delta_{i\phi}^T \delta_{i\phi}$$

$$+ \sum_{i=1}^{N} \left[e_{ix,1} \cos(\bar{\phi}_i) + e_{iy,1} \sin(\bar{\phi}_i) \right] \hat{\theta}_{i1} e_{ix,2}$$

$$+ \sum_{i=1}^{N} \left([e_{ix,1} \quad e_{iy,1}] \frac{\partial R(\phi_i)}{\partial \phi_i} \begin{bmatrix} f_{1i} \\ f_{2i} \end{bmatrix} + \delta_{i\phi} \right) \hat{\theta}_{i2} e_{i\phi,2} \tag{10.23}$$

where $\delta_x = [\delta_{1x}, \ldots, \delta_{Nx}]$ with $\delta_{ix} = x_i - x_r + \rho_{ix}$ and $\delta_y = [\delta_{1y}, \ldots, \delta_{Ny}]$ with $\delta_{iy} = y_i - y_r + \rho_{iy}$. Q is defined in (10.16).

• *Step 2*: We are now at the position to derive the actual control torque τ_i. Define $\omega_{i1d} = \alpha_{i1} + \alpha_{i2}$, $\omega_{i2d} = \alpha_{i1} - \alpha_{i2}$. $z_{i,1} = \omega_{i1} - \omega_{i1d}$, $z_{i,2} = \omega_{i2} - \omega_{i2d}$. From (10.17) and the fact that $\omega_{i1} = u_{i1} + u_{i2}$ and $\omega_{i2} = u_{i1} - u_{i2}$, there is $e_{ix,2} = 0.5(z_{i,1} + z_{i,2})$ and $e_{i\phi,2} = 0.5(z_{i,1} - z_{i,2})$. Let $z_i = [z_{i,1}, z_{i,2}]^T$. Thus we have

$$z_i = \omega_i - \begin{bmatrix} \omega_{i1d} \\ \omega_{i2d} \end{bmatrix}. \tag{10.24}$$

Multiplying the derivatives of both sides of (10.24) by M_i and combining it with (10.18) and (10.19), we obtain that

$$M_i \dot{z}_i = -D_i z_i + \Phi_i^T \Theta_i + \tau_i \tag{10.25}$$

where matrix Φ_i and Θ_i are defined as

$$\Phi_i = [\chi_i, \chi_{i,j_1}, \chi_{i,j_2}, ..., \chi_{i,j_{n_i}}]^T \tag{10.26}$$

$$\Theta_i = [\vartheta_i^T, \vartheta_{i,j_1}^T, \vartheta_{i,j_2}^T, ..., \vartheta_{i,j_{n_i}}^T]^T. \tag{10.27}$$

Note j_p for $p = 1, \ldots, n_i$ are the indexes of robot i's neighboring robots (i.e. $j_p \in \mathcal{N}_i$) of which the total number is n_i. The elements in Φ_i and Θ_i are given in (10.28).

$$\vartheta_i = [c_i r_i b_i^{-1} \quad d_{i1} \quad d_{i2} \quad m_{i1} \quad m_{i2} \quad m_{i1} r_i \quad m_{i2} r_i \quad m_{i1} r_i b_i^{-1} \quad m_{i2} r_i b_i^{-1}]^T$$

$$\vartheta_{i,j_{ni}} = [m_{i1} r_j \quad m_{i2} r_j \quad m_{i1} r_j b_j^{-1} \quad m_{i2} r_j b_j^{-1}]^T$$

$$\chi_i = \begin{bmatrix} -\omega_{i2} u_{i2} & -\omega_{i1d} & 0 & -\Delta_{i11} & -\Delta_{i12} & -\Delta_{i21} & -\Delta_{i22} & -\Delta_{i31} & -\Delta_{i32} \\ \omega_{i1} u_{i1} & 0 & -\omega_{i2d} & -\Delta_{i12} & -\Delta_{i11} & -\Delta_{i22} & -\Delta_{i21} & -\Delta_{i32} & -\Delta_{i31} \end{bmatrix}$$

$$\chi_{ij} = \begin{bmatrix} -\Delta_{ij11} & -\Delta_{ij12} & -\Delta_{ij21} & -\Delta_{ij22} \\ -\Delta_{ij12} & -\Delta_{ij11} & -\Delta_{ij22} & -\Delta_{ij21} \end{bmatrix}$$

$$\Delta_{i1k} = \frac{\partial \omega_{ikd}}{\partial \rho_{ix}} \dot{\rho}_{ix} + \frac{\partial \omega_{ikd}}{\partial \dot{\rho}_{ix}} \ddot{\rho}_{ix} + \frac{\partial \omega_{ikd}}{\partial \rho_{iy}} \dot{\rho}_{iy} + \frac{\partial \omega_{ikd}}{\partial \dot{\rho}_{iy}} \ddot{\rho}_{iy} + \frac{\partial \omega_{ikd}}{\partial f_{rx}} \ddot{f}_{rx} + \frac{\partial \omega_{ikd}}{\partial f_{ry}} \ddot{f}_{ry}$$

$$+ \frac{\partial \omega_{ikd}}{\partial \phi_r} \ddot{\phi}_r + \frac{\partial \omega_{ikd}}{\partial \hat{\theta}_{i1}} \dot{\hat{\theta}}_{i1} + \frac{\partial \omega_{ikd}}{\partial \hat{\theta}_{i2}} \dot{\hat{\theta}}_{i2} + \frac{\partial \omega_{ikd}}{\partial \hat{w}_{rx,i}} \dot{\hat{w}}_{rx,i} + \frac{\partial \omega_{ikd}}{\partial \hat{w}_{ry,i}} \dot{\hat{w}}_{ry,i}$$

$$\Delta_{i2k} = \frac{\partial \omega_{ikd}}{\partial \bar{x}_i}(\cos(\bar{\phi}_i) u_{i1}) + \frac{\partial \omega_{ikd}}{\partial \bar{y}_i}(\sin(\bar{\phi}_i) u_{i1}), \quad \Delta_{i3k} = \frac{\partial \omega_{ikd}}{\partial \bar{\phi}_i} u_{i2}$$

$$\Delta_{ij1k} = \frac{\partial \omega_{ikd}}{\partial \bar{x}_j}(\cos(\bar{\phi}_j) u_{j1}) + \frac{\partial \omega_{ikd}}{\partial \bar{y}_j}(\sin(\bar{\phi}_j) u_{j1}), \quad \Delta_{ij2k} = \frac{\partial \omega_{ikd}}{\partial \bar{\phi}_j} u_{j2}, \quad \text{for } k = 1, 2.$$

$$\tag{10.28}$$

Remark 10.5. Θ_i in (10.27) is a vector of unknown parameters involved in the ith robot dynamic subsystem, in which ϑ_i is the local unknown parameters, while ϑ_{i,j_p} is the coupled uncertainties related to the unknown parameters in robot j_p's dynamics if $a_{ij_p} = 1$. Thus online estimates of ϑ_{i,j_p}, i.e. $\hat{\vartheta}_{i,j_p}$, will be introduced in designing the torques for robot i.

Introduce the estimate $\hat{\Theta}_i$ for unknown parameter vector Θ_i. Then the local control torque and adaptive law are designed as

$$\tau_i = -K_i z_i - \Phi_i^T \hat{\Theta}_i - 0.5 \Xi_i \tag{10.29}$$

$$\dot{\hat{\Theta}}_i = \Gamma_i \Phi_i z_i \tag{10.30}$$

where $\Xi_i = [\Xi_{i,1}, \Xi_{i,2}]^T$ and

$$\Xi_{i,1} = [e_{ix,1} \quad e_{iy,1}] \left(\begin{bmatrix} \cos(\bar{\phi}_i) \\ \sin(\bar{\phi}_i) \end{bmatrix} \hat{\theta}_{i1} + \frac{\partial R(\phi_i)}{\partial \phi_i} \begin{bmatrix} f_{1i} \\ f_{2i} \end{bmatrix} \right.$$

$$\left. \times \hat{\theta}_{i2} \right) + \delta_{i\phi} \hat{\theta}_{i2}$$

$$\Xi_{i,2} = [e_{ix,1} \quad e_{iy,1}] \left(\begin{bmatrix} \cos(\bar{\phi}_i) \\ \sin(\bar{\phi}_i) \end{bmatrix} \hat{\theta}_{i1} - \frac{\partial R(\phi_i)}{\partial \phi_i} \begin{bmatrix} f_{1i} \\ f_{2i} \end{bmatrix} \right)$$

$$\left. \times \hat{\theta}_{i2} \right) - \delta_{i\phi} \hat{\theta}_{i2}. \tag{10.31}$$

Choose the Lyapunov function for the overall system as

$$V_2 = V_1 + \frac{1}{2} \left(z_i^T M_i z_i + \tilde{\Theta}_i^T \Gamma_i^{-1} \tilde{\Theta}_i \right) \tag{10.32}$$

where Γ_i is a symmetric and positive definite matrix and $\tilde{\Theta}_i = \Theta_i - \hat{\Theta}_i$. From (10.23) and (10.29), (10.30), we obtain that

$$\dot{V}_1(t) \le -\frac{k_1}{2} \left(\delta_x^T Q \delta_x + \delta_y^T Q \delta_y \right) - k_2 \delta_{i\phi}^T \delta_{i\phi}$$

$$- z_i^T (K_i + D_i) z_i. \tag{10.33}$$

The main results in this section are formally presented in the following theorem.

Theorem 10.6. *Consider the closed-loop adaptive system consisting of N nonholonomic mobile robots (10.1)-(10.2), the control torques (10.29) and parameter estimators (10.21) and (10.30) under Assumptions 10.1-10.5. The formation errors for each robot are ensured to satisfy that*

$$\lim_{t \to \infty} \bar{x}_i(t) - x_r(t) = -\rho_{ix} + \sqrt{2} \varepsilon_{i1} \tag{10.34}$$

$$\lim_{t \to \infty} \bar{y}_i(t) - y_r(t) = -\rho_{iy} + \sqrt{2} \varepsilon_{i1} \tag{10.35}$$

$$\lim_{t \to \infty} \bar{\phi}_i(t) - \phi_r(t) = \varepsilon_{i2}. \tag{10.36}$$

Proof: From (10.33), it can be shown that δ_{ix}, δ_{iy} and $\delta_{i\phi}$ will converge to zero asymptotically. This indicates that $\lim\limits_{t \to \infty} [x_i(t) - x_r(t)] = -\rho_{ix}$, $\lim\limits_{t \to \infty} [y_i(t) - y_r(t)] = -\rho_{iy}$ and $\lim\limits_{t \to \infty} [\phi_i(t) - \phi_r(t)] = 0$.

From (10.4), (10.5) and (10.7), we obtain that

$$\|(x_i - \bar{x}_i, y_i - \bar{y}_i)\| \leq \sqrt{2\varepsilon_1^2}, \quad |\phi_i - \bar{\phi}_i| \leq \varepsilon_2. \tag{10.37}$$

It then follows that

$$
\begin{aligned}
|\bar{x}_i + \rho_{ix} - x_r| &\leq |\bar{x}_i - x_i| + |x_i + \rho_{ix} - x_r| \\
|\bar{y}_i + \rho_{iy} - y_r| &\leq |\bar{y}_i - y_i| + |y_i + \rho_{iy} - y_r| \\
|\bar{\phi}_i - \phi_r| &\leq |\bar{\phi}_i - \phi_i| + |\phi_i - \phi_r|.
\end{aligned}
\tag{10.38}
$$

Since $x_i + \rho_{ix} - x_r$, $y_i + \rho_{iy} - y_r$ and $\phi_i - \phi_r$ will converge to zero asymptotically, (10.34)-(10.36) hold. As discussed in Remark 10.5, by properly adjusting ε_{i1} and ε_{i2}, the formation errors of the overall system can be made as small as desired. $\qquad\square$

10.4 Simulation Results

We now use four mobile robots to demonstrate the effectiveness of the controllers. The communication topology is given in Fig.10.2. The reference trajectory is given as follows. $x_r(t) = t$, $y_r(t) = 10\sin(0.1t)$. $\rho_{1x} = 3$, $\rho_{2x} = 3$, $\rho_{3x} = 6$, $\rho_{4x} = 6$, $\rho_{1y} = 0$, $\rho_{2y} = 3$, $\rho_{3y} = 0$, $\rho_{1y} = 3$. The parameters of the robots are chosen as: $b_i = 0.75$, $d_i = 0.3$, $r_i = 0.25$, $m_{ci} = 10$, $m_{wi} = 1$, $I_{ci} = 5.6$, $I_{wi} = 0.005$, $I_{mi} = 0.0025$, $d_{i1} = d_{i2} = 5$. The control parameters are chosen as: $\varepsilon_{i1} = 0.1$, $\varepsilon_{i2} = 0.1$, $k_1 = 2$, $k_2 = 2$, $\gamma_{\theta_{i1}} = \gamma_{\theta_{i2}} = 5$, $\gamma_{r_i} = 4$, $K_i = 2I$, $\Gamma_i = 4I$, and parameter ϵ for projection is chosen as $\epsilon = 0.1$. θ_{i1} and θ_{i2} are assumed to be in $[0.15, 0.4]$ and $[0.1, 0.3]$ respectively. The robot position is shown in Fig.10.3 and the orientation error $\delta_{i\phi}$ is shown in Fig.10.4. Clearly, these results are consistent with those stated in Theorem 10.6 and therefore illustrate our theoretical findings. Fig.10.5 shows the distances of the four robots.

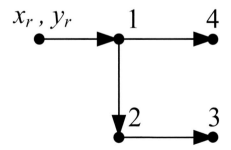

Fig. 10.2. Communication topology of the four mobile robots and the reference x_r and y_r are only available to robot 1.

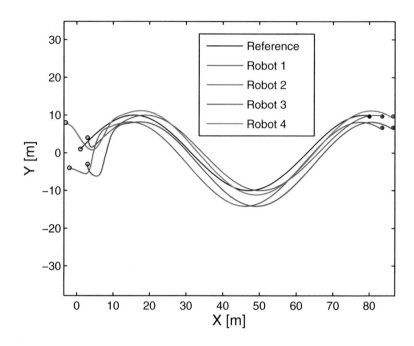

Fig. 10.3. The positions of the four mobile robots in X-Y plane.

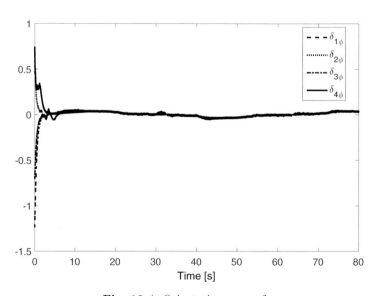

Fig. 10.4. Orientation errors $\delta_{i\phi}$

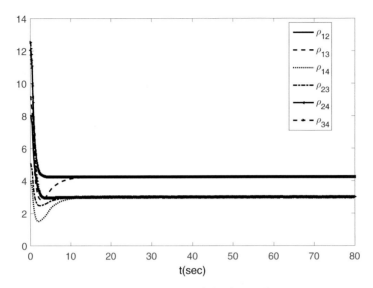

Fig. 10.5. Distances of the four robots.

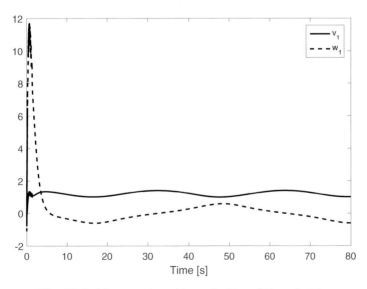

Fig. 10.6. Linear and angular velocities of the robot 1.

10.5 Conclusion

In this chapter, we have investigated the output formation problem for a collection of mobile robots. We assume that only part of the mobile robots can obtain the exact information of the desired trajectory directly. By adopting the backstepping technique, distributed adaptive control laws are designed based on the information collected within neighboring areas. It is shown that all signals in the closed-loop system are bounded and the asymptotically formation for all the mobile robots' outputs can be ensured. The simulation results show the effectiveness of the proposed control approach. Moreover, in contrast to currently available scheme in which the information exchange of online parameter estimates among linked subsystems is required, the condition can be relaxed by introducing additional estimates to account for the uncertainties in the neighbors' dynamics. Thus the desired results are achieved with less transmission burden.

1

Adaptive Flocking Control of Multiple Nonholonomic Mobile Robots

This chapter investigates the flocking problem of multiple nonholonomic mobile robots under the situations that all parameters of the robots are unknown and the robots are of limited communication ranges. Backstepping technique based distributed control schemes are proposed by introducing a p-time differential function. This function is embedded into a potential function to design control laws for flocking. The controllers guarantee that no collisions between any two robots occur, and no switching of controllers is needed. The size of the flock is bounded and the robots will go to an equilibrium set as time goes to infinity. Simulations illustrate the results.

1.1 Introduction

Over the past few years cooperative control of multiple agents has received a lot of attention from both control and robot communities due to the vast applications such as search, rescue, surveillance and cooperative transportation, etc. Cooperative control of multiple robot agents is a challenging problem because each agent's behavior is affected by its neighbors, and the whole group works as a team. Formation control of a group of agents is to control the positions of the agents such that they perform certain tasks such as stabilization or tracking desired references. The research works on formation control [100, 101] can be roughly divided into three categories: behavior based, virtual structure and leader follower. In behavior based approach [6] several desired behaviors such as collision avoidance, formation keeping and target seeking are prescribed for each robot, and a scheme to switch or interpolate between alternative behaviors is designed by weighing the relative importance of each behavior. The difficulty of this approach is that the mathematical formalization is complicated. The virtual structure [96, 11] approach considers the formation as a single rigid virtual structure and the agent trajectory is generated in a centralized manner based on states of all agents. In the leader follower approach [98] one or more agents are designated as leaders and the

rest are designated as followers. The leader tracks a desired trajectory and t
followers track the leader to maintain a desired distance and orientation to t
leader. The followers implement controllers using only the state of themselv
and the leader.

In [102], a provably-stable flocking algorithm is proposed, in which a
output vector is produced by distributed filters based on position informatie
alone but not velocity information. In [103], a novel individual-based alig
ment and repulsion algorithm is proposed such that each individual repels i
sufficiently close neighbors and aligns to the average velocity of its neighbo
with moderate distances. In [104], a connectivity-preserving flocking algorith
with a bounded potential function is proposed by combining the ideas of cc
lective potential functions and velocity consensus without assuming that tl
communication topology retains its connectivity frequently enough and tl
potential function provides an infinite force during the evolution of agents. I
[105], a distributed controller for flocking of N mobile agents with an ellipt
cal shape and with limited communication ranges is proposed. All the mode
considered in above literature are simple second-order dynamic agents. I
[13], a flocking control for mobile robots is considered, but collision avoidanc
among the robots is omitted. Also the mobile robots are not considered at th
kinematic level, and the parameters are assumed to be known. So far, flockin
control of mobile robots at the torque level with unknown system parameter
has not been investigated yet.

11.2 Problem Formulation

11.2.1 Robot Dynamics

We consider a group of N two-wheeled mobile robots as shown in Fig.11.1
which can be described by the following dynamic models

$$\dot{\eta}_i = J(\eta_i)\omega_i \tag{11.1}$$

$$M_i\dot{\omega}_i + C_i(\dot{\eta}_i)\omega_i + D_i\omega_i = \tau_i \tag{11.2}$$

where $\eta_i = (\bar{x}_i, \bar{y}_i, \bar{\phi}_i)$ denotes the position and orientation of the robot, $\omega_i = (\omega_{i1}, \omega_{i2})^T$ denote the angular velocities of the left and right wheels, $\tau_i = (\tau_{i1}, \tau_{i2})^T$ represents the control torques applied to the wheels, and M_i is
a symmetric, positive definite inertia matrix, $C_i(\dot{\eta}_i)$ is the centripetal and
coriolis matrix, D_i denotes the surface friction. Matrices $J(\eta_i)$, M_i, $C_i(\dot{\eta}_i)$
and D_i are defined as the same as those in [99].

In these expressions, b_i is the half width of the mobile robot and r_i is the
radius of the wheel, d_i is the distance between the center of the mass, P_c, of
the robot and the middle point between the left and right wheels, I_{ci}, I_{wi} and
I_{mi} are the moment of inertia of the body about the vertical axis through P_c,
the wheel with a motor about the wheel axis, and the wheel with a motor
about the diameter, respectively. The positive constants d_{ik}, $k = 1, 2$, are the
damping coefficients.

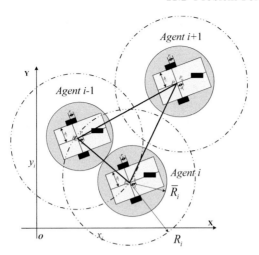

Fig. 11.1. General framework of of mobile robot flocking. Robot $i-1$ and robot i are within communication range of each other, while robot $i+1$ is not.

11.2.2 Robot Dynamics Transformation

In order to solve the underactuated problem of the mobile robot, a coordinate change is proposed based on the transverse function approach [15] as follows

$$\begin{bmatrix} x_i \\ y_i \end{bmatrix} = \begin{bmatrix} \bar{x}_i \\ \bar{y}_i \end{bmatrix} + R(\phi_i) \begin{bmatrix} f_{1i}(\alpha_i) \\ f_{2i}(\alpha_i) \end{bmatrix}$$
$$\phi_i = \bar{\phi}_i - f_{3i}(\alpha_i) \tag{11.3}$$

where $R(\phi_i) = [\cos(\phi_i), -\sin(\phi_i); \sin(\phi_i), \cos(\phi_i)]$ and $f_{li}(\alpha_i)$, $l = 1, 2, 3$ are functions of α_i which will be determined later. Computing the derivatives on both sides of x_i, y_i, ϕ_i, we have

$$\begin{bmatrix} \dot{x}_i \\ \dot{y}_i \end{bmatrix} = Q_i \begin{bmatrix} r_i u_{i1} \\ \dot{\alpha}_i \end{bmatrix} + R'(\phi_i) \begin{bmatrix} f_{1i}(\alpha_i) \\ f_{2i}(\alpha_i) \end{bmatrix} (r_i b_i^{-1} u_{i2} - f'_{3i}(\alpha_i)\dot{\alpha}_i)$$
$$\dot{\phi}_i = r_i b_i^{-1} u_{i2} - f'_{3i}\dot{\alpha}_i \tag{11.4}$$

where $\dot{\alpha}_i$ is deemed as auxiliary manipulated variable,

$$Q_i = \left[\begin{bmatrix} \cos(\bar{\phi}_i) \\ \sin(\bar{\phi}_i) \end{bmatrix} \quad R(\phi_i) \begin{bmatrix} f'_{1i}(\alpha_i) \\ f'_{2i}(\alpha_i) \end{bmatrix} \right]$$

and $u_{i1} = 0.5(\omega_{i1} + \omega_{i2})$, $u_{i2} = 0.5(\omega_{i1} - \omega_{i2})$, $f'_{li} = \dfrac{\partial f_{li}(\alpha_i)}{\partial \alpha_i}$, $l = 1, 2, 3$ and $R'(\phi_i) = \dfrac{\partial R(\phi_i)}{\partial \phi_i}$. $f_{li}(\alpha_i)$ are chosen such that Q_i is invertible for all $\bar{\phi}_i \in \Re$ and $\alpha_i \in \Re$ as follows

$$f_{1i}(\alpha_i) = \varepsilon_{1i}\sin(\alpha_i)\frac{\sin(f_{3i})}{f_{3i}}, \quad f_{2i}(\alpha_i) = \varepsilon_{1i}\sin(\alpha_i)\frac{1-\cos(f_{3i})}{f_{3i}},$$

$$f_{3i}(\alpha_i) = \varepsilon_{2i}\cos(\alpha_i) \tag{11.5}$$

where ε_{1i} and ε_{2i} are small constants selected to satisfy that

$$\varepsilon_{1i} > 0, \quad 0 < \varepsilon_{2i} < \frac{\pi}{2}.$$

It follows that

$$|f_{1i}| < \varepsilon_{1i}, \quad |f_{2i}| < \varepsilon_{1i}, \quad |f_{3i}| < \varepsilon_{2i} \tag{11.6}$$

and

$$\det(Q_i) = \frac{\varepsilon_{1i}\varepsilon_{2i}}{(\varepsilon_{2i}\cos(\alpha_i))^2}(\cos(\varepsilon_{2i}\cos(\alpha_i)) - 1)$$

$$\leq -\frac{\varepsilon_{1i}}{\varepsilon_{2i}}(1 - \cos(\varepsilon_{2i})) < 0. \tag{11.7}$$

11.2.3 A p-Time Differential Step Function

A p-time differential step-like functions will be used to avoid collisions between any two robots.

Definition [105]: A scalar function $h(x, a, b)$ which is shown in Fig.11.2 is called p-time differential step function if it possesses the following properties:

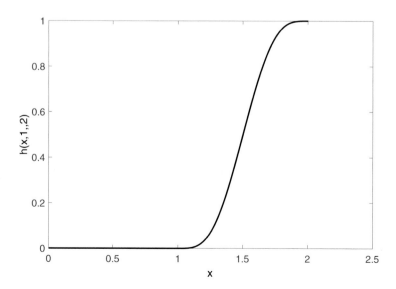

Fig. 11.2. A schematic figure of h.

1) $h(x, a, b) = 0, \forall -\infty < x \le a$
2) $h(x, a, b) = 1, \forall b \le x < \infty$
3) $0 < h(x, a, b) < 1, \forall a < x < b$
4) $h(x, a, b)$ is p-time differentiable

where p is a positive integer, $x \in \Re$, a and b are constants such that $a < b$.

Lemma 11.1. *[105]: Let the scalar function $h(x, a, b)$ be defined as*

$$h(x, a, b) = \frac{\int_a^x f(\varsigma - a) f(b - \varsigma) d\varsigma}{\int_a^b f(\varsigma - a) f(b - \varsigma) d\varsigma} \tag{11.8}$$

with the function $f(y)$ being defined as

$$f(y) = \begin{cases} 0 & \text{if } y \le 0 \\ g(y) & \text{if } y > 0 \end{cases} \tag{11.9}$$

where the function $g(y)$ has the following properties:
1) $g(\varsigma - a) g(b - \varsigma) > 0, \forall \varsigma \in (a, b)$
2) $g(y)$ is p-time differentiable
3) $\lim\limits_{y \to 0^+} \frac{\partial^k g(y)}{\partial y^k} = 0, \quad k = 1, ...p - 1.$
Then $h(x, a, b)$ is a p-time differential step function.

Proof: See [105].

An example of $g(y)$ is $g(y) = y^p$. Moreover, if $g(y)$ is chosen as $g(y) = e^{-1/y}$, then $h(x, a, b)$ is a smooth step function.

11.2.4 Formation Control Problem

In order to design a flocking control scheme for a group of mobile robots, we need to specify the goal for the group. The framework of flocking is shown in Fig. 11.1. First we impose the following assumption on the communication and initial conditions between the robots and the flocking rendezvous trajectory.

Assumption 11.1 *Robots i and j have circular communication areas, which are centered at P_{ic} and P_{jc} with radii R_i and R_j respectively. Also robots i and j have physical safety areas with radii $\bar{R}_i$ and $\bar{R}_j$. The radii R_i and R_j are sufficiently large such that $R_i > \bar{R}_i + \bar{R}_j$, $i, j = 1, 2...N$, $i \ne j$.*

Assumption 11.2 *Robot i broadcasts its state η_i in its communication area. Moreover, robot i can receive the state η_j broadcast by robot j if robot j is within the communication area of robot i, i.e. $j \in \mathcal{N}_i$ with $\mathcal{N}_i$ denoting the set of robots that are within the communication area of robot i.*

Assumption 11.3 *At the initial time, any two robots in the group have safe distances with each other. More concretely, there exists a strictly positive constant ε_1 such that*

$$\| q_i(0) - q_j(0) \| - (\bar{R}_i + \bar{R}_j) \geq \varepsilon_1 \tag{11.10}$$

for all i and j, $i \neq j$, where $q_i = [x_i, y_i]^T$.

Assumption 11.4 *The flocking rendezvous trajectory $\eta_{od} = [x_{od}, y_{od}, \phi_{od}]^T$ for all the robots to track has a bounded derivative $\dot{\eta}_{od}$ and is available to all the robots.*

Assumption 11.5 *The dimensional terms r_i and b_i are unknown, but are in known compact sets.*

Flocking Control Objective: Under Assumptions 11.1-11.5, design control input τ_i for robot i such that collision between any two robots can be avoided. Also the flocking of all the robots to the rendezvous trajectory η_{od} can be achieved with bounded errors and flocking size.

11.3 Control Design

11.3.1 Potential Function

We now construct pairwise potential functions that will be used in the Lyapunov function for control design. Let $\mathcal{D}_{ij}$ be the general distance of the centers of robots i and j, i.e. $\mathcal{D}_{ij} = \| q_i - q_j \|$ and Ω_{ij} be a pairwise potential function of λ_{ij} where λ_{ij} is defined as

$$\lambda_{ij} = \frac{1}{2}\left(\sqrt{1 + \mathcal{D}_{ij}^2} - \sqrt{1 + \mathcal{D}_{ijs}^2}\right). \tag{11.11}$$

$\mathcal{D}_{ijs}$ denotes the safe distance of robots i and j in the flocking configuration, i.e., $\mathcal{D}_{ijs} \geq \bar{R}_i + \bar{R}_j + 2\varepsilon_{1i}$, where ε_{1i} is the positive constant, which is shown in Fig.11.3. In addition, let $\mathcal{D}_{ijd}$ be the demanding distance of robots i and j in the flocking configuration and let $\mathcal{D}_{ijm}$ be the greatest lower bound of $\mathcal{D}_{ij}$ when robots i and j are within their communication ranges. We assume that $\mathcal{D}_{ijs} < \mathcal{D}_{ijd} < \mathcal{D}_{ijm}$.

Ω_{ij} should have the following properties:
1) $\Omega_{ij} = k_{ij}$, $\Omega'_{ij} = 0$, $\Omega''_{ij} = 0$, $\forall \lambda_{ij} \in [\lambda_{ijm}, \infty)$
2) $\Omega_{ij} \geq 0$, $\forall \lambda_{ij} \in (0, \lambda_{ijm})$
3) $\lim_{\lambda_{ij} \to 0} \Omega_{ij} = \infty$, $\lim_{\lambda_{ij} \to 0} \Omega'_{ij} = -\infty$
4) Ω_{ij} is at least three times differentiable.
5) Ω_{ij} has a unique minimum value at $\lambda_{ij} = \lambda_{ijd}$.
where λ_{ijd} and λ_{ijm} are λ_{ij} defined in (10) with $\mathcal{D}_{ij}$ replaced by $\mathcal{D}_{ijd}$ and $\mathcal{D}_{ijm}$, respectively. k_{ij} is a positive constant to be chosen later.

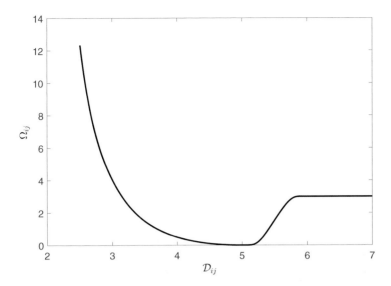

Fig. 11.3. A schematic figure of D_{ij}.

Remark 11.2. Property 1 means that Ω_{ij} is zero when robots i and j are outside their communication ranges. Property 2 means Ω_{ij} is positive when robots i and j are within their communication ranges. Property 3 means $\Omega_{ij} = \infty$ when a collision between robots i and j happens. Property 4 implies that we can use the backstepping technique to design the flocking control.

Based on the properties, we design a potential Ω_{ij} function as follows

$$\Omega_{ij} = \mu_{ij}\frac{1 - h_{ij}(\lambda_{ij}, a_{ij}, b_{ij})}{\lambda_{ij}} + k_{ij}h_{ij}(\lambda_{ij}, a_{ij}, b_{ij}) \tag{11.12}$$

where $h_{ij}(\lambda_{ij}, a_{ij}, b_{ij})$ is a p-time differentiable step function with $p \geq 3$ and $g(y)$ defined as $g(y) = y^p$. k_{ij} is a positive constant to be chosen. μ_{ij} is defined as

$$\mu_{ij} = \frac{1}{2}\left(\sqrt{1 + \mathcal{D}_{ij}^2} - \sqrt{1 + \mathcal{D}_{ijd}^2}\right)^2.$$

Constants a_{ij} and b_{ij} are chosen such that

$$a_{ij} = \lambda_{ijd}, \quad b_{ij} \leq \lambda_{ijm}$$

Lemma 11.3. *[105]: Ω_{ij} satisfies Properties 1) - 5).*

Proof: See [105].

We construct the potential function for the whole group of robots as

$$\Omega = \frac{1}{2} \sum_{i=1}^{N-1} \sum_{j=i+1}^{N} \Omega_{ij}. \tag{11.13}$$

The derivative of Ω is

$$\dot{\Omega} = \sum_{i=1}^{N-1} \sum_{j=i+1}^{N} \frac{\partial \Omega_{ij}}{\partial \mathcal{D}_{ij}^2} q_{ij}^T \dot{q}_{ij} = \sum_{i=1}^{N-1} \sum_{j=i+1}^{N} \Omega'_{ij} q_{ij}^T (\dot{q}_i - \dot{q}_{od})$$

$$- \sum_{i=1}^{N-1} \sum_{j=i+1}^{N} \Omega'_{ij} q_{ij}^T (\dot{q}_j - \dot{q}_{od}) \tag{11.14}$$

where $q_{ij} = q_i - q_j$ and $\Omega'_{ij} = \dfrac{\partial \Omega_{ij}}{\partial \mathcal{D}_{ij}^2}$.

11.3.2 Flocking Control Design

In this section, we will design the controls for τ_i in two steps. In the first step we will design the virtual controls for u_{i1} and u_{i2}, together with $\dot{\alpha}_i$ such that the control objective is achieved. In the second step we will design controls for τ_i such that the virtual control errors approach zero.

Step 1:

System (11.4) and (11.2) form a strict feedback system. Therefore we apply a Lyapunov function and the backstepping technique to design controls τ_i. The Lyapunov function is constructed as follows:

$$V_0 = \Omega + \frac{1}{2} \sum_{i=1}^{N} \left(k_1 \parallel q_i - q_{od} \parallel^2 + k_2 (\phi_i - \phi_{od})^2 \right) \tag{11.15}$$

where $q_{od} = [x_{od} \ y_{od}]^T$, k_1 and k_2 are positive constants. Taking the derivative of (11.15) results in

$$\dot{V}_0 = \sum_{i=1}^{N} \left[[\Omega_{ix} \ \Omega_{iy}] \left(Q_i \begin{bmatrix} r_i u_{i1} \\ \dot{\alpha}_i \end{bmatrix} + R'(\phi_i) \begin{bmatrix} f_{1i}(\alpha_i) \\ f_{2i}(\alpha_i) \end{bmatrix} \right) \right.$$

$$\times (r_i b_i^{-1} u_{i2} - f'_{3i}(\alpha_i)\dot{\alpha}_i) - \dot{q}_{od})$$

$$\left. + \Omega_{i\phi} \left(r_i b_i^{-1} u_{i2} - f'_{3i}\dot{\alpha}_i - \dot{\phi}_{od} \right) \right] \tag{11.16}$$

where

$$\begin{bmatrix} \Omega_{ix} \\ \Omega_{iy} \end{bmatrix} = k_1 (q_i - q_{od}) + \sum_{j=i+1}^{N} \Omega'_{ij} q_{ij}^T - \sum_{j=1}^{i-1} \Omega'_{ij} q_{ij}^T$$

$$\Omega_{i\phi} = k_2 (\phi_i - \phi_{od}).$$

Let $\theta_{i1} = r_i$, $\theta_{i2} = r_i b_i^{-1}$, $\hat{\theta}_{i1}$ and $\hat{\theta}_{i2}$ be the estimates of θ_{i1} and θ_{i2} respectively and define $\tilde{\theta}_{i1} = \theta_{i1} - \hat{\theta}_{i1}$, $\tilde{\theta}_{i2} = \theta_{i2} - \hat{\theta}_{i2}$. Let u_{i1d} and u_{i2d} be the virtual controls of u_{i1} and u_{i2} respectively and define the following errors

$$u_{i1e} = u_{i1} - u_{i1d}, \quad u_{i2e} = u_{i2} - u_{i2d}. \tag{11.17}$$

From (11.16), we design the virtual controls u_{i1d}, u_{i2d} and additional control $\dot{\alpha}_i$ as follows

$$\begin{bmatrix} u_{i1d} \\ \dot{\alpha}_i \end{bmatrix} = Q_i^{-1} \Theta_i \Big(-c_1 \psi(\Omega_i) - R'(\phi_i) \begin{bmatrix} f_{1i}(\alpha_i) \\ f_{2i}(\alpha_i) \end{bmatrix}$$

$$(-c_2 \psi(\Omega_{i\phi}) + \dot{\phi}_{od}) + \dot{q}_{od} \Big)$$

$$u_{i2d} = \frac{1}{\hat{\theta}_{i2}} \Big(-c_2 \psi(\Omega_{i\phi}) + f'_{3i} \dot{\alpha}_i + \dot{\phi}_{od} \Big) \tag{11.18}$$

where c_1 and c_2 are positive constants, $\psi(\Omega_i) = [\psi(\Omega_{ix}) \ \psi(\Omega_{iy})]^T$, and $\Theta_i = [\frac{1}{\hat{\theta}_{i1}} \ 0; 0 \ 1]$. The function $\psi(x)$ is a scalar, differentiable, bounded function that satisfies

1) $|\psi(x)| < M_1$
2) $\psi(x) = 0$ if $x = 0$, $x\psi(x) > 0$ if $x \neq 0$
3) $\psi(-x) = -\psi(x)$, $(x - y)[\psi(x) - \psi(y)] \geq 0$
4) $|\frac{\psi(x)}{x}| \leq M_2$, $|\frac{\partial \psi(x)}{\partial x}| \leq M_3$, $\frac{\partial \psi(x)}{\partial x}\Big|_{x=0} = 1$

for all $x \in \Re$, $y \in \Re$, where M_1, M_2 and M_3 are positive constants. Some functions satisfying these properties are $\arctan(x)$ and $\tanh(x)$. Substituting (11.18) into (11.16) results in

$$\dot{V}_0 = -c_1 \sum_{i=1}^{N} \Omega_{ix} \psi(\Omega_{ix}) - c_1 \sum_{i=1}^{N} \Omega_{iy} \psi(\Omega_{iy}) - c_2 \sum_{i=1}^{N} \Omega_{i\phi} \psi(\Omega_{i\phi})$$

$$+ \sum_{i=1}^{N} \Big[\varrho_{i\theta 1} \tilde{\theta}_{i1} + \varrho_{i\theta 2} \tilde{\theta}_{i2} + \varrho_{iu1} u_{i1e} + \varrho_{iu2} u_{i2e} \Big] \tag{11.19}$$

where

$$\varrho_{i\theta 1} = [\Omega_{ix} \ \Omega_{ix}] Q_i [1 \ 0]^T u_{i1}$$

$$\varrho_{i\theta 2} = \Big([\Omega_{ix} \ \Omega_{ix}] R'(\phi_i) [f_{1i}(\alpha_i) \ f_{2i}(\alpha_i)]^T + \Omega_{i\phi} \Big) u_{i2}$$

$$\varrho_{iu1} = [\Omega_{ix} \ \Omega_{ix}] Q_i [1 \ 0]^T \hat{\theta}_{i1}$$

$$\varrho_{iu2} = \Big([\Omega_{ix} \ \Omega_{ix}] R'(\phi_i) [f_{1i}(\alpha_i) \ f_{2i}(\alpha_i)]^T + \Omega_{i\phi} \Big) \hat{\theta}_{i2}.$$

Consider another Lyapunov function

$$V_1 = V_0 + \frac{1}{2} \sum_{i=1}^{N} (\gamma_1 \tilde{\theta}_{i1}^2 + \gamma_2 \tilde{\theta}_{i2}^2)$$

where γ_1 and γ_2 are positive constants. We get

$$
\begin{aligned}
\dot{V}_1 = &-c_1 \sum_{i=1}^{N} \Omega_{ix}\psi(\Omega_{ix}) - c_1 \sum_{i=1}^{N} \Omega_{iy}\psi(\Omega_{iy}) - c_2 \sum_{i=1}^{N} \Omega_{i\phi}\psi(\Omega_{i\phi}) \\
&+ \sum_{i=1}^{N} \left[\varrho_{iu1}u_{i1e} + \varrho_{iu2}u_{i2e} \right] + \sum_{i=1}^{N} \left[\gamma_1\tilde{\theta}_{i1}(\dot{\hat{\theta}}_{i1} + \gamma_1^{-1}\varrho_{i\theta 1}) \right. \\
&\left. + \gamma_2\tilde{\theta}_{i2}(\dot{\hat{\theta}}_{i1} + \gamma_2^{-1}\varrho_{i\theta 2}) \right].
\end{aligned}
$$

The parameter update laws are designed to be

$$
\begin{aligned}
\dot{\hat{\theta}}_{i1} &= \text{Proj}\left(-\gamma_1^{-1}\varrho_{i\theta 1},\ \hat{\theta}_{i1} \right) \\
\dot{\hat{\theta}}_{i2} &= \text{Proj}\left(-\gamma_2^{-1}\varrho_{i\theta 2},\ \hat{\theta}_{i2} \right).
\end{aligned}
\tag{11.20}
$$

The operator "Proj" is a continuous projection algorithm adopted in our case as follows

$$
\text{Proj}(a, \hat{b}) = \begin{cases} a & b_{\min} + \zeta < \hat{b} < b_{\max} - \zeta \\ a + 0.5[1 + a^2] & \hat{b} < b_{\min} + \zeta \\ a - 0.5[1 + a^2] & \hat{b} > b_{\max} - \zeta \end{cases}
\tag{11.21}
$$

where ζ is a small positive constant. According to Assumption 11.5, r_i and b_i are in known compact sets. So for r_i and $r_i b_i^{-1}$, as denoted by b in (11.21), there exists $b_{\max}, b_{\min}$ and a small constant ζ such that $b_{\min} + \zeta < b < b_{\max} - \zeta$. When $\hat{b} \le b_{\min} + \zeta$, we have $\dot{\hat{b}} \ge 0$, and this law will prevent $\hat{b}$ from getting smaller than $b_{\min}$. Also we get $\tilde{b} = \hat{b} - b < 0$ and $\dot{\hat{b}} - a > 0$, therefore we have $\tilde{b}(\dot{\hat{b}} - a) < 0$. Similarly, when $\hat{b} \ge b_{\max} - \zeta$, we have $\dot{\hat{b}} \le 0$, and this law will prevent $\hat{m}$ from getting greater than $b_{\max}$. Also we get $\tilde{b} = \hat{b} - b > 0$ and $\dot{\hat{b}} - a < 0$.

Based on the above analysis, we have

$$
\begin{aligned}
\dot{V}_1 \le &-c_1 \sum_{i=1}^{N} \Omega_{ix}\psi(\Omega_{ix}) - c_1 \sum_{i=1}^{N} \Omega_{iy}\psi(\Omega_{iy}) - c_2 \sum_{i=1}^{N} \Omega_{i\phi}\psi(\Omega_{i\phi}) \\
&+ \sum_{i=1}^{N} \left[\varrho_{iu1}u_{i1e} + \varrho_{iu2}u_{i2e} \right].
\end{aligned}
\tag{11.22}
$$

Step 2:

Define $w_{i1e} = w_{i1} - w_{i1d}$, $w_{i2e} = w_{i2} - w_{i2d}$, where $w_{i1d} = u_{i1d} + u_{i2d}$, $w_{i2d} = u_{i1d} - u_{i2d}$. From (11.17) we get $u_{i1e} = 0.5(w_{i1e} + w_{i2e})$ and $u_{i2e} = 0.5(w_{i1e} - w_{i2e})$. Let $w_{ie} = w_i - w_{id}$ where $w_{ie} = [w_{i1e}\ w_{i2e}]^T$ and $w_{id} = [w_{i1d}\ w_{i2d}]^T$. Differentiating both sides of $w_{ie} = w_i - w_{id}$, multiplying the results by M_i and using (11.2), we get

$$M_i \dot{\omega}_{ie} = -D_i \omega_{ie} + \Phi_i \Theta_i + \tau_i \qquad (11.23)$$

where matrix Φ_i and Θ_i are defined as

$$\Theta_i = [c_i r_i b_i^{-1} \; d_{i1} \; d_{i2} \; m_{i1} \; m_{i2} \; m_{i1} r_i \; m_{i2} r_i \; m_{i1} r_i b_i^{-1} \; m_{i2} r_i b_i^{-1}]^T$$

$$\Phi_i = \begin{bmatrix} -\omega_{i2} u_{i2} & -\omega_{i1d} & 0 & -\Delta_{i11} & -\Delta_{i12} & -\Delta_{i21} & -\Delta_{i22} & -\Delta_{i31} & -\Delta_{i32} \\ \omega_{i1} u_{i2} & 0 & -\omega_{i2d} & -\Delta_{i12} & -\Delta_{i11} & -\Delta_{i22} & -\Delta_{i21} & -\Delta_{i32} & -\Delta_{i31} \end{bmatrix}$$

$$\Delta_{i1k} = \frac{\partial \omega_{ikd}}{\partial x_{od}} \dot{x}_{od} + \frac{\partial \omega_{ikd}}{\partial \dot{x}_{od}} \ddot{x}_{od} + \frac{\partial \omega_{ikd}}{\partial y_{od}} \dot{y}_{od} + \frac{\partial \omega_{ikd}}{\partial \dot{y}_{od}} \ddot{y}_{od} + \frac{\partial \omega_{ikd}}{\partial \phi_{od}} \dot{\phi}_{od}$$

$$+ \frac{\partial \omega_{ikd}}{\partial \dot{\phi}_{od}} \ddot{\phi}_{od} + \frac{\partial \omega_{ikd}}{\partial \alpha_i} \dot{\alpha}_i + \frac{\partial \omega_{ikd}}{\partial \hat{\theta}_{i1}} \dot{\hat{\theta}}_{i1} + \frac{\partial \omega_{ikd}}{\partial \hat{\theta}_{i2}} \dot{\hat{\theta}}_{i2}$$

$$\Delta_{i2k} = \frac{\partial \omega_{ikd}}{\partial x_i} [1 \; 0] Q_i [1 \; 0]^T u_{i1} +$$

$$\sum_{m=1, m \neq i}^{N} \frac{\partial \omega_{ikd}}{\partial y_m} [1 \; 0] Q_m [0 \; 1]^T u_{m1}, \; \Delta_{i3k} = \frac{\partial \omega_{ikd}}{\partial \phi_i} u_{i2} + \frac{\partial \omega_{ikd}}{\partial \bar{\phi}_i} u_{i2}$$

where $k = 1, 2$.

To design the actual control τ_i, define a new Lyapunov function

$$V_2(t) = V_1(t) + 0.5(\omega_{ie}^T M_i \omega_{ie} + \tilde{\Theta}_i^T \Lambda_i^{-1} \tilde{\Theta}_i) \qquad (11.24)$$

where adaptation gain Λ_i is a symmetric and positive definite matrix, $\tilde{\Theta}_i = \Theta_i - \hat{\Theta}_i$ with $\hat{\Theta}_i$ being the estimate of vector Θ_i. Differentiating $V_2(t)$, then we can choose the control and parameter update laws as

$$\tau_i = -K_i \omega_{ie} - \Phi_i \hat{\Theta}_i - 0.5 \Xi$$

$$\dot{\hat{\Theta}}_i = \Lambda_i \Phi_i^T \omega_{ie} \qquad (11.25)$$

where $K_i = K_i^T > 0$ and $\Xi = \left(\varrho_{iu1} + \varrho_{iu2}, \varrho_{iu1} - \varrho_{iu2} \right)^T$ such that

$$\dot{V}_2(t) \leq -c_1 \sum_{i=1}^{N} \Omega_{ix} \psi(\Omega_{ix}) - c_1 \sum_{i=1}^{N} \Omega_{iy} \psi(\Omega_{iy}) - c_2 \sum_{i=1}^{N} \Omega_{i\phi} \psi(\Omega_{i\phi})$$

$$- \tilde{\omega}_i^T (K_i + D_i) \tilde{\omega}_i. \qquad (11.26)$$

Now we state our main result in the following theorem.

Theorem 11.4. *Under Assumptions 11.1- 11.5, the proposed control input τ_i and parameter update laws $\dot{\hat{\theta}}_{i1}$, $\dot{\hat{\theta}}_{i2}$ and $\dot{\hat{\Theta}}_i$ in (11.25) achieve the flocking control objective. In particular, the following results hold:*

1) There is no collision between any two robots.

2) The relative distance between each robot and the flocking rendezvous trajectory q_{od} is bounded, i.e., $\|q_i - q_{od}\| \le c_3$, $|\phi_i - \phi_{od}| < c_4$ where c_3 and c_4 are positive constants depending on the initial conditions.

3) The flocking size $F_L(t) = \sum\limits_{i=1}^{N-1} \sum\limits_{j=i+1}^{N} D_{ij}$ is bounded, i.e.,

$$\sup F_L(t) < c_5 \tag{11.27}$$

where c_5 is a positive constant.

Proof:

1) From (11.25) it is clear that the control τ_i is continuous.

From (11.26) we get $\dot{V}_2(t) \le 0$. Integrating both sides of $\dot{V}_2(t) \le 0$ from 0 to t results in $V_2(t) \le V_2(0)$ for all $t \ge 0$. Therefore $V_2(t)$ is bounded by a positive constant that depends on the initial condition. From the properties of Ω_{ij}, there is no collision between any two robots, because Ω_{ij} will be infinity if a collision happens.

2) From the definition of $V_2(t)$ we get that $\Omega_{ij}(t)$, $\|q_i(t) - q_{od}(t)\|$ and $\phi_i - \phi_{od}$ are bounded. Since $V_2(t) \le V_2(0)$, $\|q_i - q_{od}\| \le c_3$, $|\phi_i - \phi_{od}| < c_4$. c_3 and c_4 are depended on the initial condition $V_2(0)$.

3) From (11.18) u_{i1d}, u_{i2d} and $\dot{\alpha}_i$ are bounded. Since ω_{i1e} and ω_{i2e} are bounded, therefore ω_{i1} and ω_{i2} are bounded. From (11.20) and (11.25) τ_i, $\dot{\hat{\theta}}_{i1}$, $\dot{\hat{\theta}}_{i2}$ and $\dot{\hat{\Theta}}_i$ are bounded. So it can be easily checked that $\dot{\Omega}_{ix}$, $\dot{\Omega}_{iy}$ and $\dot{\Omega}_{i\phi}$ are bounded. Therefore from Barbalat's Lemma we have

$$\lim_{t \to \infty} \left(\Omega_{ix}, \Omega_{iy}, \Omega_{i\phi} \right) = 0. \tag{11.28}$$

Let $q_{ic} = [x_{ic} \ y_{ic} \ \phi_{ic}]^T$, $0 < i \le N$ be the equilibrium state of robot i as t go to infinity. From (11.28) we have

$$k_1(q_{ic} - q_{od}) + \sum_{j=i+1}^{N} \Omega'_{ijc} q_{ijc}^T - \sum_{j=1}^{i-1} \Omega'_{ijc} q_{ijc}^T = 0 \tag{11.29}$$

where $q_{ijc} = q_{ic} - q_{jc}$ and Ω_{ijc} are Ω_{ij} with q_{ij} replaced by q_{ijc}. From (11.15) and (11.26) $\|q_{ic} - q_{od}\|$ is bounded by a positive constant that depends on the initial condition $V_0(0)$. Thus $\sum\limits_{j=i+1}^{N} \Omega'_{ijc} q_{ijc}^T - \sum\limits_{j=1}^{i-1} \Omega'_{ijc} q_{ijc}^T$ is bounded by a positive constant. From (11.28)

$$k_1 q_{ijc} + \left(\sum_{l=i+1}^{N} \Omega'_{ilc} q_{ilc}^T - \sum_{l=1}^{i-1} \Omega'_{ilc} q_{ilc}^T \right)$$

$$- \left(\sum_{k=j+1}^{N} \Omega'_{jkc} q_{jkc}^T - \sum_{k=1}^{j-1} \Omega'_{jkc} q_{jkc}^T \right) = 0. \tag{11.30}$$

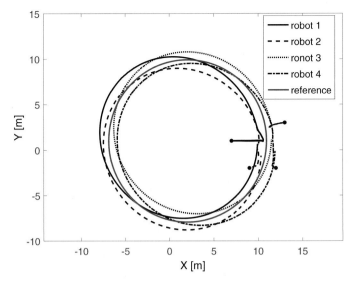

Fig. 11.4. Positions of robots in the X-Y plane.

So from (11.29) $\|q_{ijc}\|$ is bounded by a positive constant that depends on initial condition. So the flocking size F_L is also bounded by a positive constant, i.e.,

$$\sup F_L(t) < c_5$$

where c_5 is a positive constant that depends on initial conditions.

11.4 Simulation Results

We simulate the flocking control of four identical mobile robots to illustrate the effectiveness of our proposed control scheme. The parameters of the robots are chosen as: $b_i = 0.75$, $d_i = 0.3$, $r_i = 0.25$, $m_{ci} = 10$, $m_{wi} = 1$, $I_{ci} = 5.6$, $I_{wi} = 0.005$, $I_{mi} = 0.0025$, $d_{i1} = d_{i2} = 5$, $R_i = 1$. The initial values of $\hat{\theta}_{i1}$, $\hat{\theta}_{i2}$ and $\hat{\Theta}_i$ are taken as 60% of their true values. $\mathcal{D}_{ijs} = 0.5$ for $1 \le i, j \le 4$, $\mathcal{D}_{12d} = 2$, $\mathcal{D}_{13d} = 2\sqrt{2}$, $\mathcal{D}_{14d} = 2$, $\mathcal{D}_{23d} = 2$, $\mathcal{D}_{24d} = 2\sqrt{2}$, $\mathcal{D}_{34d} = 2$. The reference trajectory is a circle. The control gains are chosen as $k_{i1} = 12$, $k_{i2} = 12$, $K_i = 4I$, $\Lambda_i = 3I$. Simulation results are shown in Fig.11.4 and Fig.11.5. It can be seen from the figures that there is no collision between any two robots, and the size of the flocking is bounded. Fig.11.6 and Fig.11.7 show the velocities and torques of Robot 1.

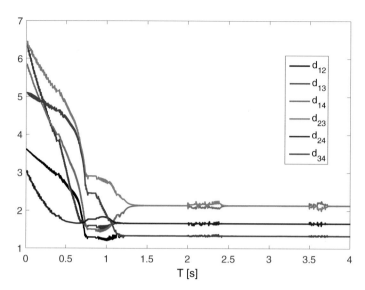

Fig. 11.5. Distance d_{ij} between any two robots i and j $(i \neq j)$.

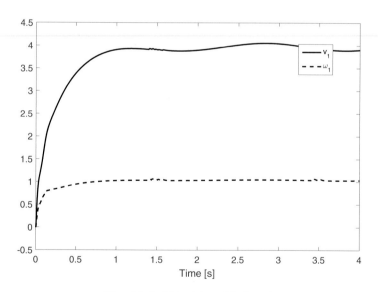

Fig. 11.6. Velocities of Robot 1.

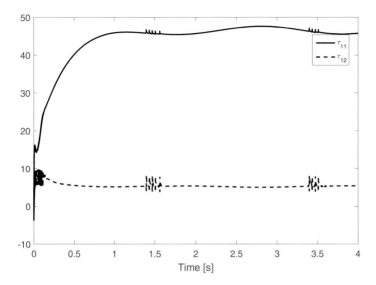

Fig. 11.7. Torque of Robot 1.

11.5 Conclusion

In this chapter we present a constructive method to design flocking controllers for a group of N mobile robots with unknown parameters. The design is based on a potential function, backstepping technique and Lyapunov functions. The designed continuous control guarantees that there is no collision between any two robots despite the limited communication ranges and the size of the flock is bounded.

References

1. R. W. Brockett. Asymptotic stability and feedback stabilization, *Differential Geometric Control Theory*, Birkhauser, Boston, page 181-191, 1983.
2. C. Samson. Time-varying feedback stabilization of car-like wheeled mobile robots. *International Journal of Robotics Research*, 12(1):55-64, 1993.
3. J. Guldner and V. I. Utkin. Stabilization of nonholonomic mobile robots using Lyapunov function for navigation and sliding mode control. *Proceedings of the 33rd IEEE Conference on Decision and Control*, page 2967-2972, 1994.
4. I. Kolmanovsky and N. McClamroch. Development in nonholonomic control problems. *IEEE Control System Magazine*, 15:20-36, 1995.
5. O. J. Sordalen and O. Egeland. Exponential stabilization of nonholonomic chained systems, *IEEE Trans. on Automatic Control*, 40:35-49, 1995.
6. Z.-P. Jiang and H. Nijmeijer. Tracking control of mobile robots: A case study in Backstepping. *Automatica*, 33(7):1393-1399, 1997.
7. T. Fukao, H. Nakagawa, N. Adachi. Adaptive tracking control of a nonholonomic mobile robot, *IEEE Trans. Robotics and Automation*, 16:609-615, 2000.
8. M. Krstic, I. Kanellakopoulos, P. Kokotovic. *Nonlinear and Adaptive Control Design*. John Wiley and Sons, 1995
9. K.-D. Do, Z. P. Jiang, J. Pan. Simultaneous tracking and stabilization of mobile robots: An adaptive approach. *IEEE Trans. Automatic Control*, 49(7):1147-1152, 2004.
10. K.-D. Do, Z.-P. Jiang and J. Pan. A global output-feedback controller for simultaneous tracking and stabilization of unicycle-type mobile robots, *IEEE Trans. on Robotics and Automation*, 20:589-594, 2004.
11. K.-D. Do and J. Pan. Nonlinear formation control of unicycle-type mobile robots, *Robotics and Autonomous Systems*, 55(3): 191-204, 2007.
12. K.-D. Do. Formation tracking control of unicycle-type mobile robots with limited sensing ranges, *IEEE Transactions on Control Systems Technology*, 16(3): 527-538, 2008.
13. W. Dong. Flocking of multiple mobile robots based on backstepping, *IEEE Transactions on Systems, Man, and Cybernetics–Part B: Cybernetics*, 41(2):414-424, 2011.
14. P. Morin and C. Samson. Contorl of nonholonomic systems based on transverse function approach. *IEEE Trans. Robotics*, 25:1058-1073, 2009.

15. P. Morin and C. Samson. Practical stabilization of driftless systems on lie group: the traverse function approach, *IEEE Transactions on Automatic Control*, 48(9): 1496-1508, 2003.
16. K. Pettersen and H. Nijmeijer. Underactuated ship tracking control: theory and experiments, *Int. J. Control*, 74(14):1435-1446, 2001.
17. Z.-P. Jiang. Global tracking control of underactuated ships by Lyapunov's direct method, *Automatica*, 38:301-309, 2002.
18. K.-D. Do, Z.-P. Jiang, J. Pan. Underactuated ship global tracking under relaxed conditions, *IEEE Trans. Automatic Control*, 47(9):1529-1536, 2002.
19. K.-D. Do, Z.-P. Jiang, J. Pan. Universal controllers for stabilization and tracking of underactuated ships, *Systems Control Letters*, 47(4):299-317, 2002.
20. A. Behal, D. M. Dawson, W. E. Dixon, and Y. Fang. Tracking and regulation control of an underactuated surface vessel with nonintegrable dynamics, *IEEE Trans. Automat. Contr.*, 47(3):495-500, 2002.
21. K.-D. Do, J. Pan. Global robust adaptive path following of underactuated ships, *Automatica*, 42(10):1713-1722, 2006.
22. K.-D. Do. Practical control of underactuated ships, *Ocean Engineering*, 37:1111-1119, 2010.
23. Y. Fang, E. Zergeroglub, M. de Queirozc, D. Dawson. Global output feedback control of dynamically positioned surface vessels: an adaptive control approach, *Mechatronics* 14(4):341-356, 2004.
24. K.-D. Do and J. Pan. Global tracking control of underactuated ships with nonzero off-diagonal terms in their system matrices. *Automatica* 41(1):87-95, 2005.
25. K.-D. Do, Z.-P. Jiang, J. Pan. Global partial-state feedback and output-feedback tracking controllers for underactuated ships, *Systems Control Letters* 54(10):1015-1036, 2005.
26. V. Sankaranarayanan and D. Mahindrakar. Control of a class of underactuated mechanical systems using sliding modes. *IEEE Trans. Robotics*, 25(2):459-467, 2009.
27. R. Ortega, M.W. Spong, F. Gomez-Estern, G. Blankenstein. Stabilization of a class of underactuated mechanical systems via interconnection and damping assignment, *IEEE Transactions on Automatic Control*, 47(8):1218-1233, 2002.
28. R. Olfati-Saber. Normal forms for underactuated mechanical systems with symmetry, *IEEE Transactions on Automatic Control*, 47(2):305-308, 2002.
29. M. Zhang and T. Tarn. A hybrid switching control strategy for nonlinear and underactuated mechanical systems, *IEEE Transactions on Automatic Control*, 48(10):1777-1782, 2003.
30. M. Reyhanoglu, A. van der Schaft, N.H. McClamroch, I. Kolmanovsky. Dynamics and control of a class of underactuated mechanical systems, *IEEE Transactions on Automatic Control*, 44(9):1663-1671, 1999.
31. T. Lee and Z.-P. Jiang. Uniform asymtotic stability of nonlinear swicthed systems with an application to mobile robots, *IEEE Trans. Automatic Control*, 53:1235-1252, 2008.
32. C. Samson and K. Ait-Abderrahim. Feedback control of a nonholonomic wheeled cart in cartesian space, *Proceeding of IEEE International Conference on Robotics Automation*, Sacramento, page 1136-1141, 1991.
33. T. Lee and K.-T. Song, Tracking control of unicycle-modeled mobile robots using a saturation feedback controller, *IEEE Trans. Control Systems Technology*, 9(2):305-318, 2001.

34. N. Perez-Arancibia, T.-C. Tsao and J. S. Gibson. Saturation-induced instability and its avoidance in adaptive control of hard disk drives, *IEEE Trans. Control Systems Technology*, 18:368-382, 2010.
35. V. Kapila and K. Grigoriadis. *Actuator Saturation Control*, CRC Press, 2002.
36. Z.-P. Jiang, E. Lefeber and H. Nijmeijer. Saturation stabilization and tracking of a nonholonomic mobile robot, *Systems Control Letters*, 42:327-332, 2001.
37. P. Lin and Y. Jia. Distributed rotating formation control of multi-agent systems, *Systems Control Letters*, 59(10):587-595, 2010.
38. W. Ren. Trajectory tracking for unmanned air vehicles with velocity and heading rate constraints, *IEEE Trans. Control Systems Technology*, 12(5):706-716, 2004.
39. W. Ren, Distributed cooperative attitude synchronization and tracking for multiple rigid bodies, *IEEE Trans. Control Systems Technology*, 18(2):383-392, 2010.
40. Z. Chen and H.-T. Zhang. No-beacon collective circular motion of jointly connected multi-agents, *Automatica*, 47(9):1929-1937, 2011.
41. H.-T. Zhang, Z. Chen, L. Yan and W. Yu, Applications of collective circular motion control to multirobot systems, *IEEE Trans. Control Systems Technology*, 21(4):1416-1422, 2012.
42. E. Lefeber, K. Pettersen and H. Nijmerjer. Tracking control of an underactuated ship, *IEEE Transactions on control systems technology*, 11(1):52-61, 2003.
43. Y. Hong, Y. Xu and J. Huang. Finite-time control for robot manipulators. *Systems & Control Letters*, 46:243-253. 2002.
44. M. Rosier. Homogeneous Lyapunov function for homogeneous continuous vector field. *Systems & Control Letters*, 19:467-473, 1992.
45. K. Pettersen and O. Egeland. Time-varying exponential stabilization of the position and attitude of an underactuated autonomous underwater vehicle, *IEEE Trans. on Automatic Control*, 44(1):112-115, 1999.
46. K. Pettersen and E. Lefeber. Way-point tracking control of ships. *Proceedings of Conference on Decision and Control*, page 940-945, 2001.
47. D. Chwa. Global Tracking control of underactuated ships with input and velocity constraints using dynamic surface control method, *IEEE Transactions on control systems technology*, 19(6):1357-1370, 2011.
48. C. Wen J. Zhou, Z. Liu and H. Su. Robust adaptive control of uncertain nonlinear systems in the presence of input saturation and external disturbance, *IEEE Transactions on Automatic Control*, 56(7):1672-1678, 2011.
49. S. Ge, Z. Wang and T.H. Lee. Adaptive stabilization of uncertain nonholonomic systems by state and output feedback. *Automatica*, 39(8):1451-1460, 2003.
50. J. Luo and P. Tsiotras. Exponentially convergent control laws for nonholonomic systems in power form. *Systems & Control Letters*, 35:87-95, 1998.
51. R. Fierro and F. Lewis. Control of a nonholonomic mobile robot: backstepping kinematics into dynamics. *Journal of Robotic Systems*, 14(3):149-163, 1997.
52. M. Egerstedt, X. Hu and A. Stotsky. Control of mobile platforms using a virtual vehicle approach. *IEEE Trans. on Automatic Control*, 46:1777-1782, 2001.
53. W. Dong, W. Huo, S.K. Tso and W.L. Xu. Tracking control of uncertain dynamic nonholonomic system and its application to wheeled mobile robots, *IEEE Trans. on Robotics and Automation*, 16(6):870-874, 2000.
54. A. Loria. Global tracking control of one-degree-of-freedom Euler-Lagrange systems without velocity measurement. *Eur. J. Control*, 2(2):144-151, 1996.

55. H. Nijmeijer and T. I. Fossen. *New Directions in Nonlinear Observer Design*, Springer-Verlag, London, U.K. 1999.
56. Z.-P. Jiang. Lyapunov design of global state and output feedback tracker for nonholonomic control systems. *Int. J. Control*, 73(9):744-761, 2000.
57. K.-W. Lee and H. K. Khalil. Adaptive output feedback control of robot manipulators using high-gain observer. *Int. J. Control*, 67(6):869-886, 1997.
58. H. Khalil. Adaptive output feedback control of nonlinear systems represented by input-output models. *IEEE Trans. Automatic Control*, 41:177-188, 1996.
59. A.N. Tikhonov. Systems of differential equations containing small parameters in the derivatives. *Mat. Sborn.* 31(73):575-586, 1952.
60. H. Ashrafiuon, K. Muske, L. McNinch and R. Soltan. Sliding-mode tracking control of surface vessels. *IEEE Trans. Industrial Electronics*, 55:4004-4012, 2008.
61. J. Ghommam and F. Mnif. Coordinated path-following control for a group of underactuated surface vessels, *IEEE Trans. Industrial Electranics* , 56(10):3951-3963, 2009.
62. C. Yan, G. Venayagamoorthy and K. Corzine. AIS-based coordinated and adaptive control of generator excitation systems for an electric ship, *IEEE Trans. Industrial Electranics*, 59(8):3102-3112, 2012.
63. K.-D. Do. Global robust and adaptive output feedback dynamic positioning of surface ships, *IEEE Conf. on Robotics and Automation*, page 4271-4276, 2007.
64. M. Polycarpou and P. Ioannou. A robust adaptive nonlinear control design, *Automatica*, 32:423-427, 1996.
65. W. Wang and C. Wen. Adaptive actuator failure compensation control of uncertain nonlinear systems with guaranteed transient performance. *Automatica*, 46(12):2082-2091, 2010.
66. M. Khalil, *Nonlinear Systems (3rd ed.)*, Prentice-Hall, Englewood Cliffs, New Jersey, 2002
67. S. Bhat and D. Bernstein. Continuous finite-time stabilization of the translational and rotional double integrators, *IEEE Transactions on Automatic Control*, 43(11):678-682, 1998.
68. M. Reyhanoglu. Exponential stabilization of an underactuated autonomous surface vessel, *Automatica*, 32:2249-2254, 1997.
69. K. Pettersen. *Exponential stabilization of underactuated vehicles*, Ph.D. Thesis, Norwegian University of Science Technology, 1996.
70. K. Pettersen and T. Fossen. Underactuated dynamic positioning of a ship-experimental results, *IEEE Trans. Control Systems Technology*, 8:856-863, 2000.
71. R. Olfati-Saber. *Nonlinear control of underactuated mechanical systems with application to robotics and aerospace vehicles*, Ph.D. Thesis, Massachusetts Institute of Technology, 2001.
72. W. Ren and Y. Cao. *Distributed Coordination of Multi-agent Networks: Emergent Problems, Models and Issues*, Springer-Verlag, London, 2010.
73. A. Behal, D. Dawson, X. Bin, and P. Setlur. Adaptive tracking control of underactuated surface vessels, *Proc. of the Conference on Control Applications*, Mexico City, Mexico, page 645-650, 2001.
74. A. Alessandri, M. Caccia, and G. Veruggio. Fault detection of actuator faults in unmanned underwater vehicles, *Control Engineering Practice*, 7(3):357-368, 1999.

75. M. Blanke, M. Staroswiecki and N. E. Wu. Concepts and methods in fault-tolerant control, *Proceedings of the American Control Conference*, page 2606-2620, Arlington, USA, 2001.

76. F. Liao, J. Wang and G. Yang. Reliable robust flight tracking control: an LMI approach, *IEEE Trans. Control System Technology*, 10(1):76-89, 2002.

77. J. Boskovic and R. Mehra. A decentralized scheme for accommodation of multiple simultaneous actuator failures, *Proceedings of the American Control Conference*, page 5098-5103, 2002.

78. F. Chen, S. Zhang, B. Jiang and G. Tao. Multiple-model based fault detection and diagnosis for helicopter with actuator faults via quantum information technique, *J. System Control Engineering*, 228(3):182-190, 2014.

79. M. Corradini and G. Orlando. Actuator failure identification and compensation through sliding modes, *IEEE Trans. Control System Technology*, 15(1):184-190, 2007.

80. F. Chen, B. Jiang and G. Tao. An intelligent self-repairing control for nonlinear MIMO systems via adaptive sliding mode control technology, *J. Franklin Institute*, 351(1):399-411, 2014.

81. K. Veluvolu, M. Defoort, Y. Soh. High-gain observer with sliding mode for nonlinear state estimation and fault reconstruction, *J. Franklin Institute*, 351(4):1995-2014, 2014.

82. J. Ghommam, F. Mnif, A. Benali and N. Derbel. Asymptotic Backstepping Stabilization of an Underactuated Surface Vessel. *IEEE Trasactions on Control Systems Technology*, 14(6): 1150-1157, 1992.

83. J. Huang, C. Wen, W. Wang, Y. Song. Global stable tracking control of underactuated ships with input saturation, *Systems & Control Letters*, 85:1-7, 2015.

84. M. Harmouche, S. Laghrouche, Y. Chitour. Global tracking for underactuated ships with bounded feedback controllers, *Int. Journal of Control*, 87(10):2035-2043, 2014.

85. R. Patton. Fault-tolerant control systems, *Proceedings of the IFAC Symposium on Fault Detection Supervision and Safety for Technical Processes*, page 1033-1054, Brussels, Belgium, 1997.

86. N. Sarkar, T. Podder and G. Antonelli. Fault accommodating thruster force allocation of an AUV considering thruster redundancy and saturation, *IEEE Transactions on Robotics and Automation*, 18(2):223-233, 2002.

87. T. Perez and A. Donaire. Constrained control design for dynamic positioning of marine vehicles with control allocation, *Modeling, Identification and Control*, 30(2):57-70, 2009.

88. E. Omerdic and G. Roberts. Thruster fault diagnosis and accommodation for open-frame underwater vehicles, *Control Engineering Practice*, 12(12):1575-1598, 2004.

89. B. Sun, D. Zhu, and L. Sun. A tracking control method with thruster fault tolerant control for unmanned underwater vehicles, *Proceedings of the 25th Chinese Control and Decision Conference (CCDC 2013)*, page 4915-4920, Guiyang, China, 2013.

90. X. Chen and W. W. Tan. Tracking control of surface vessels via fault-tolerant adaptive backstepping interval type-2 fuzzy control, *Ocean Engineering*, 70:97-109, 2013.

91. G. Toussaint, T. Basar, and F. Bullo. Tracking for nonlinear underactuated surface vessels with generalized forces, *Proceedings of the IEEE International Conference on Control Applications*, page 355-360, Anchorage, Alaska, USA, 2000.

92. P. Morin and C.Samson. Time-varying exponential stabilization of the attitude of a rigid space craft with two controls, *Proc. of the IEEE Conf. on Decision and Control*, New Orleans, LA, page 3988-3993, 1995.

93. M. Krstic, I. Kanellakopoulos and P. Kokotovic, *Nonlinear and Adaptive Control Design*, John Wiley & Sons, New York, 1995.

94. J. Pomet and L. Praly. Adaptive nonlinear regulation: estimation from the Lyapunov equation, *IEEE Trans. Automatic Control*, 37(6):729-740, 1992.

95. C. Wang et al. Decentralized adaptive backstepping control for a class of interconnected nonlinear systems with unknown actuator failures, *Journal of the Franklin Institute*, 352:835-850, 2015.

96. M. Lewis and K.-H. Tan. High precision formation control of mobile robots using virtual structures, *Auton. Robots*, 4(4):387-403, 1997.

97. H. Zhang and F. Lewis. Adaptive cooperative tracking control of higher-order nonlinear systems with unknown dynamics, *Automatica*, 48(7):1432–1439, 2012.

98. A. Das, R. Fierro, et al. A vision based formation control framework, *IEEE Trans. Robotics and Automation*, 18(5):813-825, 2002.

99. D. Stipanovic, G. Inalhan, R. Teo and C. J. Tomlin. Decentralized overlapping control of a formation of unmanned aerial vehicles, *Automatica*, 40(8):1285-1296, 2004.

100. T. Balch and R. C. Arkin. Behavior-based formation control for multirobotteams, *IEEE Trans. Robotics and Automation*, 14(6):926-939, 1998.

101. R. Jonathan, R. Beard, and B. Young. A decentralized approach to formation maneuvers, *IEEE Trans. Robotics and Automation*, 19(6):933-941, 2003.

102. H. Su, X. Wang, G. Chen. A connectivity-preserving flocking algorithm for multi-agent systems based only on position measurements, *Int. J. Control*, 82(7):1334-1343, 2009.

103. H. Zhang, C. Zhai and Z. Chen. A general alignment repulsion algorithm for flocking of multi-agent systems, *IEEE Trans. Automatic Control*, 56(2):430-435, 2011.

104. G. Wen, Z. Duan, et al. A connectivity-preserving flocking algorithm for multi-agent dynamical systems with bounded potential function. *IET Control Theory and Applications* 6(6):813-821, 2011.

105. K.-D. Do. Flocking for multiple elliptical agents with limited communication ranges, *IEEE Trans. Automatic Control*, 27:931-942, 2011.

Index